ULF Pulsations in
the Magnetosphere

ADVANCES IN EARTH AND PLANETARY SCIENCES

Advances in Earth and Planetary Sciences 11

Supplement Issue to Journal of Geomagnetism and Geoelectricity

ULF Pulsations in the Magnetosphere

Reviews from the Special Sessions on
Geomagnetic Pulsations at XVII General Assembly of
the International Union for Geodesy and Geophysics,
Canberra, 1979, December

Edited by
D. J. Southwood

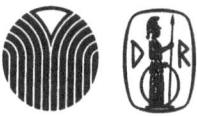

Center for Academic Publications Japan/Tokyo
D. Reidel Publishing Company/Dordrecht·Boston·London

Library of Congress Cataloging in Publication Data

DATA APPEARS ON SEPARATE CARD

ISBN-13: 978-94-009-8428-8   e-ISBN-13: 978-94-009-8426-4
DOI: 10.1007/978-94-009-8426-4

Published by Center for Academic Publications Japan, Tokyo, in co-publication with D. Reidel Publishing Company, P.O.Box 17, 3300 AA Dordrecht.

Sold and distributed in Japan, China, Korea, Taiwan, Indonesia, Cambodia, Laos, Malaysia, Philippines, Thailand, Vietnam, Burma, Pakistan, India, Bangla Desh, Sri Lanka by Center for Academic Publications Japan, 4-16, Yayoi 2-chome, Bunkyo-ku, Tokyo 113, Japan.

Sold and distributed in the U.S.A. and Canada by Kluwer Boston Inc., 190 Old Derby Street, Hingham, MA 02043, U.S.A.

Sold and distributed in all other countries by Kluwer Academic Publishers Group, P.O. Box 322, 3300 AH Dordrecht, Holland.

Preface

Physical and chemical studies of the earth and planets along with their surroundings are now developing very rapidly. As these studies are of essentially international character, many international conferences, symposia, seminars and workshops are held every year. To publish proceedings of these meetings is of course important for tracing development of various disciplines of earth and planetary sciences though publishing is getting fast to be an expensive business.

It is my pleasure to learn that the Center for Academic Publications Japan and the Japan Scientific Societies Press have agreed to undertake the publication of a series "Advances in Earth and Planetary Sciences" which should certainly become an important medium for conveying achievements of various meetings to the academic as well as non-academic scientific communities. It is planned to publish the series mostly on the basis of proceedings that appear in the Journal of Geomagnetism and Geoelectricity edited by the Society of Terrestrial Magnetism and Electricity of Japan, the Journal of Physics of the Earth by the Seismological Society of Japan and the Volcanological Society of Japan, and the Geochemical Journal by the Geochemical Society of Japan, although occasional volumes of the series will include independent proceedings.

Selection of meetings, of which the proceedings will be included in the series, will be made by the Editorial Committee for which I have the honour to work as the General Editor. I and the members of the Editorial Committee will certainly welcome any suggestions that will promote the series. Whenever the convener of a meeting related to earth and planetary sciences is in a position to have to look for a medium for publishing the proceedings please contact us.

Tsuneji Rikitake
General Editor

Introduction

It is more than twenty-five years since J. W. Dungey first suggested that the sinusoidal oscillations in the Earth's field first recorded almost a hundred years before and known as geomagnetic pulsations were due to hydromagnetic waves in the Earth's magnetosphere. The last ten years have seen a rapid growth in the understanding of the phenomena due to a happy combination of theoretical and experimental advances. Six IAGA (International Association for Geomagnetism and Aeronomy) sessions at the IUGG General Assembly in Canberra in December, 1979 were devoted to pulsation topics and formed probably the largest symposium on the subject yet held. Here we present some of the papers given at the meeting, in particular eight review papers. The reviewers were invited by an international organising committee; B. J. Fraser (Australia), R. Gendrin (France), F. Glangeaud (France), L. J. Lanzerotti (U.S.A.), R. L. McPherron (U.S.A.), V. A. Troitskaya (U.S.S.R.), which was chaired by me.

We aimed to get a balanced cross section of papers representing current international efforts. An informed reader can make his own judgement of our success by looking at the papers here. I am sad that two reviewers could not provide papers for this collection. Alain Roux provided a fascinating report on the latest results on high frequency pulsations (Pc 1 band and even above) emerging from the S-300 wave experiment on GEOS. Oleg Raspopov gave a review of Pi 2 substorm associated pulsations measured with magnetometer arrays. Anyone who attended the meeting knows that both topics are of burgeoning interest now. Apart from these omissions, and perhaps a slight weighting towards experiment, my view is that this collection does well represent the present research directions. I trust it does turn out to be useful.

D. J. Southwood
Blackett Laboratory
Imperial College
London

CONTENTS

Preface v

Introduction ....D. J. SOUTHWOOD   vii

Latitudinal and Longitudinal Variation of Pc 4, 5 Pulsations and Implications
Regarding Source Mechanisms..
.........................G. ROSTOKER, J. C. SAMSON, and J. V. OLSON   1

Observations of Pc Pulsations in the Magnetosphere: Satellite-Ground
Correlation ....S. KOKUBUN   17

Multisatellite Observations of Geomagnetic Pulsations............W. J. HUGHES   41

Substorm Associated Micropulsations at Synchronous Orbit..................
.... R. L. McPHERRON   57

Low Frequency Pulsation Generation by Energetic Particles....D. J. SOUTHWOOD   75

Solar Wind Control of Daytime, Midperiod Geomagnetic Pulsations.........
...............E. W. GREENSTADT, R. L. McPHERRON, and K. TAKAHASHI   89

Pulsation Structure in the Ionosphere Derived from Auroral Radar Data......
..........................A. D. M. WALKER and R. A. GREENWALD   111

Damping and Coupling of Long-Period Hydromagnetic Waves by the Ionosphere
....F. B. KNOX and W. ALLAN   129

The Rotation of Hydromagnetic Waves by the Ionosphere..................
...............M. K. ANDREWS, L. J. LANZEROTTI, and C. G. MACLENNAN   141

Latitudinal and Longitudinal Variation of Pc 4, 5 Pulsations and Implications Regarding Source Mechanisms

Gordon Rostoker, John C. Samson, and John V. Olson

Institute of Earth and Planetary Physics and Department of Physics,
University of Alberta, Edmonton, Alberta, Canada

(Received June 28, 1980)

Early theoretical studies of geomagnetic pulsations dealt with the oscillation of individual magnetic lines of force in the toroidal mode. In recent years it has become increasingly clear that the pulsations are organized in the azimuthal direction so that treatment of an oscillating surface of field lines may need to be considered rather than oscillations of individual field lines. In this paper evidence for the azimuthal dependence of Pc 4, 5 pulsation characteristics will be presented. In addition, the latitudinal characteristics of the ground perturbation pattern and the local time variation of these pulsations will be discussed in the context of some of the source mechanisms proposed for long period micropulsation activity.

1. Introduction

Since the organized study of geomagnetic pulsations was initiated, there has been a tendency to define two distinct categories of pulsational activity—impulsive and continuous. Present terminology introduced by Jacobs et al. (1964) provided the terms Pc for the continuous activity and Pi for impulsive activity. However, as one can see from Fig. 1, even so-called Pc activity can often appear impulsive in character, having the appearance of short-lived wave pockets whose lifetimes are sufficiently long so that the wavetrains blend together leaving the impression that the activity is continuous. The Pc 4, 5 activity discussed in this paper has precisely the characteristics noted above, and the reader should keep this in mind when we discuss the possible source mechanisms for geomagnetic pulsational activity. We merely note here that our observations relate to the Pc 4 ($6.7 < f < 22.2$ mHz) and Pc 5 ($1.7 < f < 6.7$ mHz) frequency ranges and pertain for the most part to the morning and early afternoon local time sectors.

The studies of geomagnetic pulsations have generally featured episodes of theoretical activity coming as an effort to explain new improved observation. The first major theoretical advances were made by Dungey (1954) who treated the magnetosphere as a hydromagnetic medium of infinite conductivity perturbed by small amplitude magnetic fields. His approach demanded a treatment of the linearized hydromagnetic equations which led to the wave equation

$$\partial^2 E / \partial t^2 = V_A \times V_A \times \nabla \times \nabla \times E$$

where $V_A = (B/(\mu_0 \rho)^{1/2})$ is the Alfvén velocity. This equation broke into two coupled component equations which represented the toroidal and poloidal modes. Further progress

1

at that time involved decoupling the equations by assuming axial symmetry which implies an azimuthal wave number equal to zero. As we shall see later in this paper, such an assumption is no longer tenable.

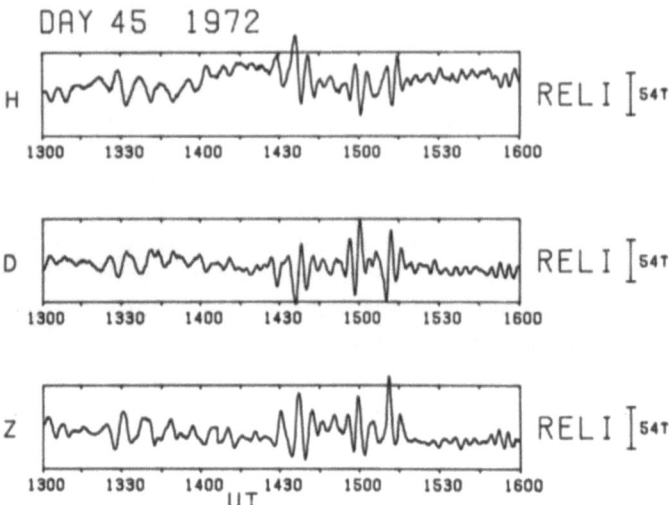

Fig. 1. Pc 5 activity typical of the auroral oval in the morning sector recorded at Fort Reliance (~70.3°N geomagnetic). Note the burstlike character of the activity (after ROSTOKER and LAM, 1978).

Dungey's approach did not particularly attempt to define the source of the energy for the hydromagnetic resonances, and it was not until the phenomonology of pulsations had been more carefully defined that a new theoretical effort was undertaken to explain how energy was supplied to the resonance region. In particular, the work of SAMSON *et al.* (1971), SAMSON (1972), and SAMSON and ROSTOKER (1972) helped to stimulate the elegant theoretical studies of CHEN and HASEGAWA (1974a, b) and SOUTHWOOD (1974) which showed how energy could be coupled into the shear Alfvén mode (viz. the waves which represent a field line resonance). At the same time some strong feeling developed that Kelvin-Helmholtz instabilities on the magnetopause boundary provided the wave energy which could be coupled into the resonance region. We shall address this question later in this paper.

Finally, we note that the new theories step away from the old hydromagnetic treatments by introducing azimuthal wave numbers different from zero. This stepped away from the concept of oscillating field lines, since it implies organized structure in the perturbation pattern normal to the magnetic field lines. We shall also discuss the implications of non-zero azimuthal wave numbers in this paper through a demonstration of the east-west and north-south scale sizes of micropulsation perturbation patterns.

2.   Azimuthal Variation of the Pc 4, 5 Pulsation Pattern

The question of whether or not pulsation signals have an azimuthal propagation velocity is not a new one, however it is only recently that the data have been adequately addressed so as to provide a definitive answer. Earlier studies by HERRON (1966) and GREEN (1976) produced confusing results because the length of the time series they analysed was such that many discrete wave pockets had their phase and amplitude characteristics averaged together. In fact, Herron noted sudden phase jumps in his data every few cycles and, as we have pointed out earlier, this suggests that the so-called continuous pulsations are not really continuous but have many of the properties of impulsive pulsations. The first fruitful studies of azimuthal phase characteristics have recently been carried out using satellite data by HUGHES et al. (1978) and using ground based data by OLSON and ROSTOKER (1978).

Hughes et al. used data from three geostationary satellites (ATS-6, SMS-1, and SMS-2) which were in close proximity to one another during February 1975. Figure 2 shows phase differences between pairs of satellites for events in the Pc 3, Pc 4, and Pc 5 frequency ranges. In the Pc 4, 5 range it is clear that the sign of the azimuthal wave number changes across local noon. The sign of the change is consistent with phase propagation away from local noon towards the dawn and dusk terminations. The wave numbers have magnitudes ranging from zero (azimuthal symmetry) to about 10 (suggestive of an azimuthal wavelength of $\sim 26,000$ km at 6.6 R_E).

OLSON and ROSTOKER (1978) used data from an east-west array of three ground

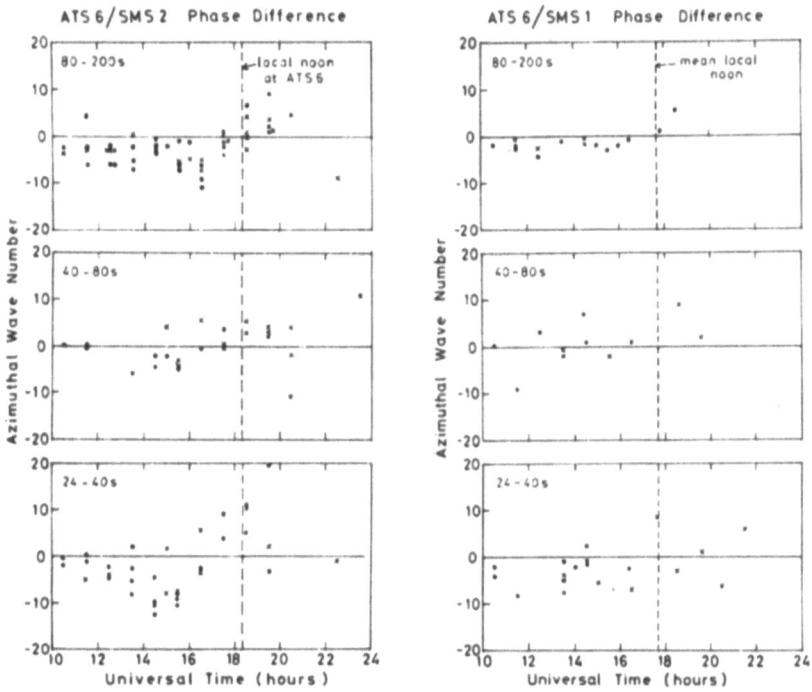

Fig. 2. Azimuthal wave numbers of Pc micropulsations in three frequency bands shown as a function of universal time (after HUGHES et al., 1978). Note change in the sign of the wave number across the approximate position of local noon.

AZIMUTHAL WAVE NUMBER

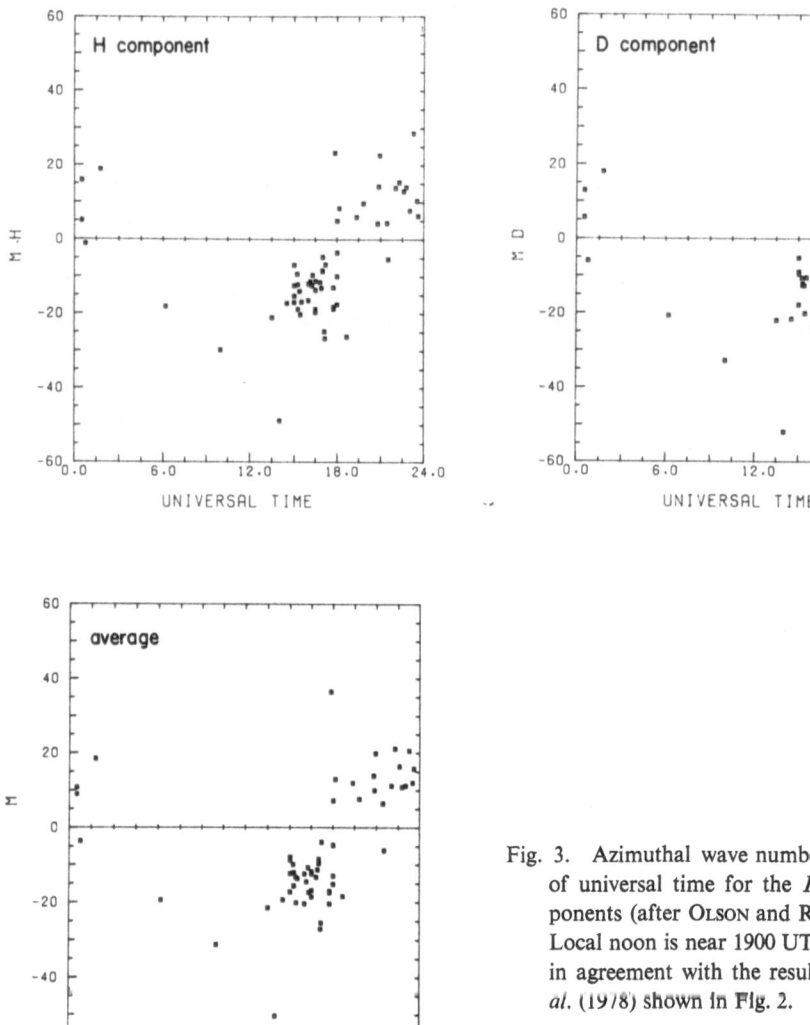

Fig. 3.   Azimuthal wave number as a function
of universal time for the *H*- and *D*-com-
ponents (after OLSON and ROSTOKER, 1978).
Local noon is near 1900 UT.   The data are
in agreement with the results of HUGHES *et
al.* (1978) shown in Fig. 2.

based magnetometers separated from each other by about 5° of longitude at ∼67.3°N
geomagnetic.   These results are shown in Figs. 3 and 4.   Figure 3 is analogous to the
results of HUGHES *et al.* (1978) shown in Fig. 2, in that it presents the azimuthal wave
number as a function of local time (UT minus 7 hours).   The results in Fig. 3 show the
azimuthal wave number obtained from *H*-component and *D*-component data as well as
data points which represent the average wave number for *H* and *D* relevant to each
event studied.   These data are consistent with the results of HUGHES *et al.* in that they
suggest that the phase propagation is away from ±1100 LT toward the terminators.

AZIMUTHAL WAVE NUMBER

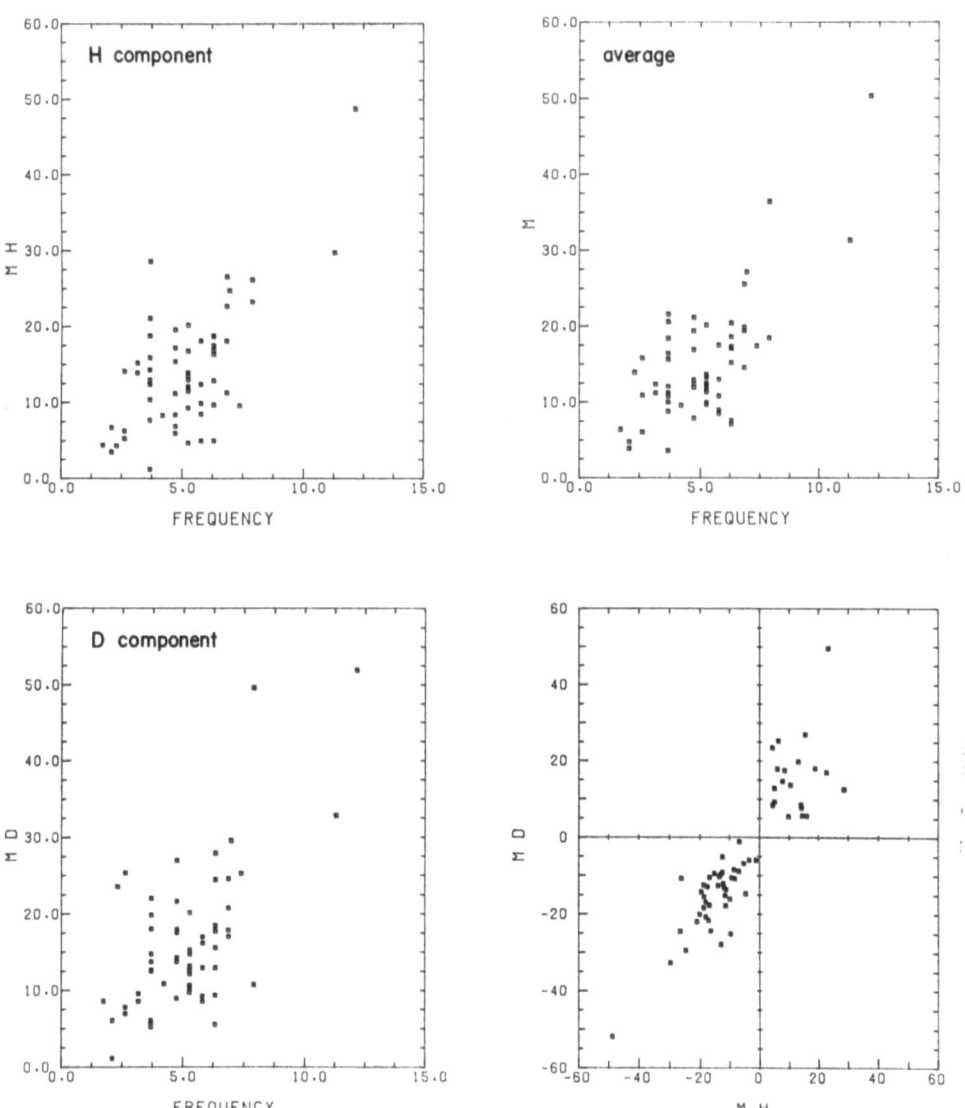

Fig. 4. Azimuthal wave number as a function of frequency for the *H*- and *D*-components and the correlation between the wave numbers for *H*- and *D*-components (after OLSON and ROSTOKER, 1978).

In addition, Fig. 4 shows a nearly linear relationship between azimuthal wave number and the frequency of the pulsations, viz.

$$m=(1.4\pm0.4)f+0.26$$

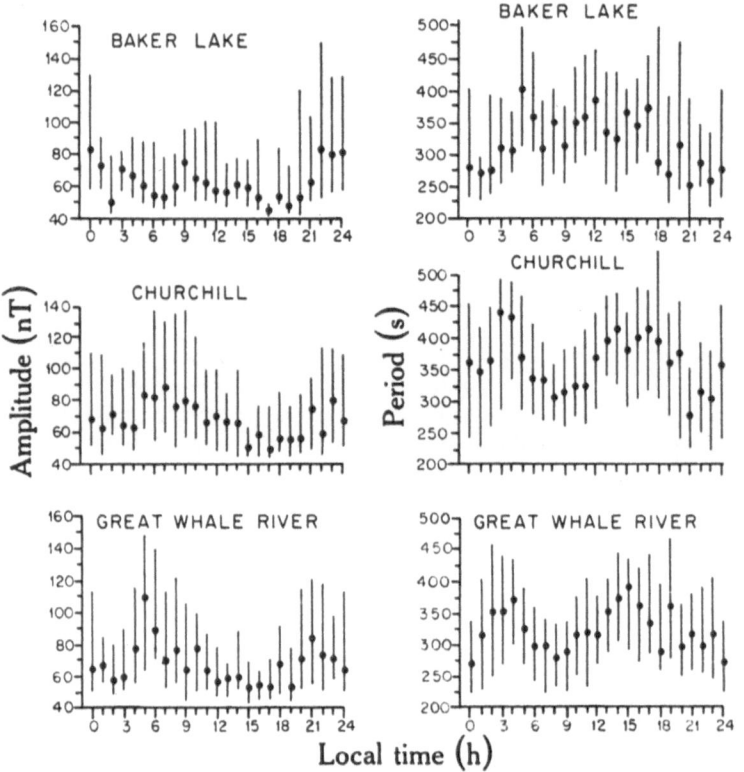

Fig. 5. Occurrence frequency of Pc 5 pulsations at a high latitude (Baker Lake), and two auroral oval (Churchill and Great Whale River) stations (after Gupta, 1976). Note the asymmetry in occurrence between the pre-noon and post-noon quadrants.

which suggests a phase velocity whose magnitude is independent of frequency and which has a value of ~ 14 km/sec at $\sim 67°$N.

On the surface, these results would appear to provide powerful support for the Kelvin-Helmholtz instability as a source of the pulsation energy. However, there are other observations of Pc 4, 5 which cause one to be rather cautious. For example, there is a strong local time asymmetry in the level of Pc 4, 5 activity with a preponderance of activity observed in the pre-noon quadrant and the activity level being clearly suppressed in the post-noon quadrant (see Fig. 5). There is no a priori reason for a Kelvin-Helmholz instability on the magnetopause to be preferentially excited on the pre-noon flank as opposed to the post-noon flank. In addition, the earlier suggestion by Kato and Utsumi (1964) among others, that the sense of polarization reversed across local noon was based on uncertain results for post-noon pulsations (see Fig. 6) where later examination indicates a confused situation where there is a slight tendency (on a statistical basis) suggesting that for most events a polarization reversal takes place. The Kelvin-Helmholtz instability is still a strong candidate as a source for Pc 4, 5 energy, however future research is needed to explain the asymmetric behaviour of pulsational activity across the noon meridian.

While phase measurements are a good measure of the azimuthal wavelength of Pc 5

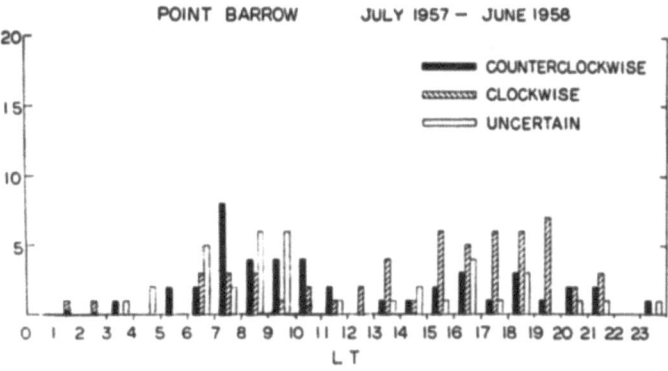

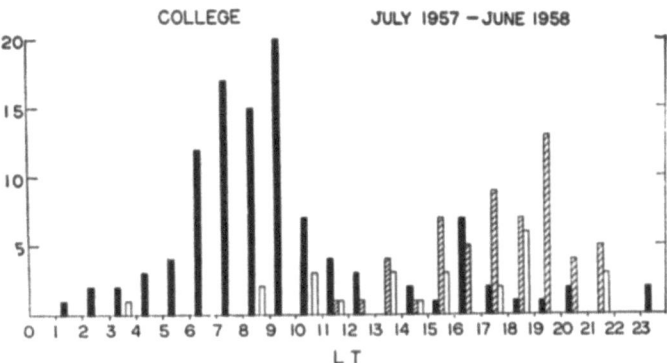

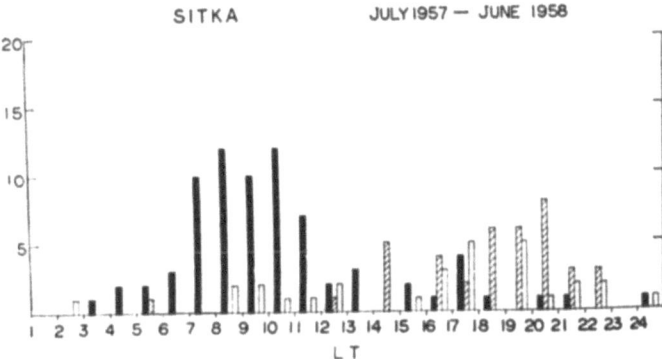

Fig. 6. Polarization in the horizontal plane for Pc 5 pulsations at a high latitude (Point Barrow), an auroral oval (College), and a sub auroral zone (Sitka) observatory (after KATO and UTSUMI, 1964). Note the confused behaviour in the sense of polarization in the post-noon sector.

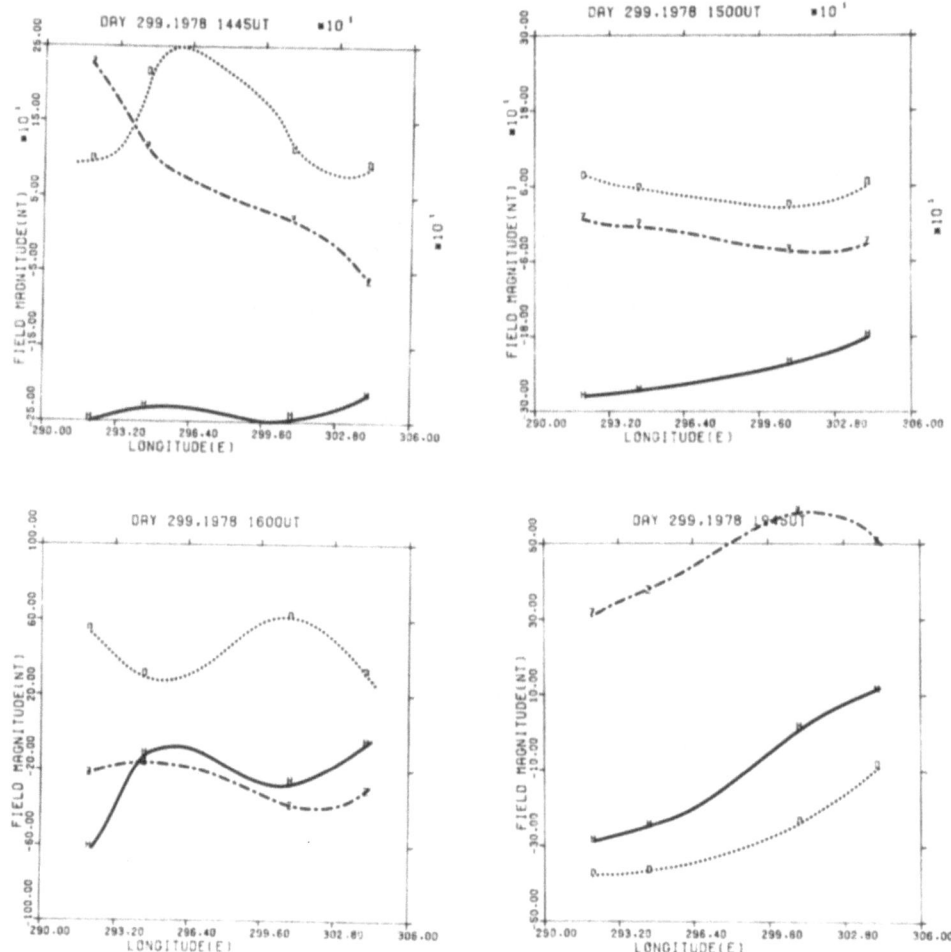

Fig. 7. Longitude profiles of the magnetic perturbation along the average auroral oval
in the morning sector showing a Ps 6 disturbance (1445 UT) and a Pc 5 event (1600 UT)
together with steady electrojet perturbation patterns (1500 UT and 1845 UT).

pulsations, it is sometimes useful to look at the amplitude variations as a function of
longitude.   Figure 7 shows several longitude profiles of the disturbance field measured
by the University of Alberta IMS magnetometer array.   The stations Fort Providence
($\sim 292.0°$E), Hay River ($\sim 294.3°$E), Fort Smith ($\sim 300.0°$E), and Uranium City (~ 304.3
°E) lie along a common latitude line of $\sim 67.4°$N.   The profile at 1445 UT shows a Ps 6
event, which involves the eastward propagation of an auroral Ω-band along the auroral
oval (KAWASAKI and ROSTOKER, 1979).   The scale size of the positive D-component spike
is clearly evident in this frame.   The profiles at 1500 UT and 1845 UT show relatively
steady electrojet perturbations, while the event shown at 1600 UT is typical of Pc 5 activity.   For this event something slightly larger than a complete wavelength is confined
within the station array.   The longitudinal wavelength for this event is $\sim 12°$ corresponding

to an azimuthal wave number of ~ 30.

3. Pc 4, 5 Pulsations and the Auroral Electrojets

While magnetopause boundary oscillations provide one possible source of Pc 4, 5 activity, recent results by LAM and ROSTOKER (1978) suggest that one might also look for a source mechanism on the field lines where the pulsations are observed. This contention is made based on the fact that it appears that all Pc 5 frequency components peak on field lines which penetrate the auroral electrojets. An example of data demonstrating this point is shown in Fig. 8, where isointensity contours of Pc 5 activity in the frequency range $1.3 < f < 2.5$ mHz are shown over the morning sector against the backdrop of the westward auroral electrojet. The fact that Pc 5 activity in this frequency range is confined to the latitudinal range occupied by the westward electrojet is clearly evident. Figure 9 shows that this trend is true for all the frequency components which make up the Pc 4, 5 spectrum. In fact, one can see in this figure some hint of the frequency-latitude dependence of the pulsations which had been theoretically predicted many years ago by DUNGEY (1954) and JACOBS and WESTPHAL (1964) among others and which was subsequently confirmed by SAMSON et al. (1971). The effect would be most obvious in the Z-component which peaks directly underneath the resonance region. Finally we show, in Fig. 10, the latitudinal

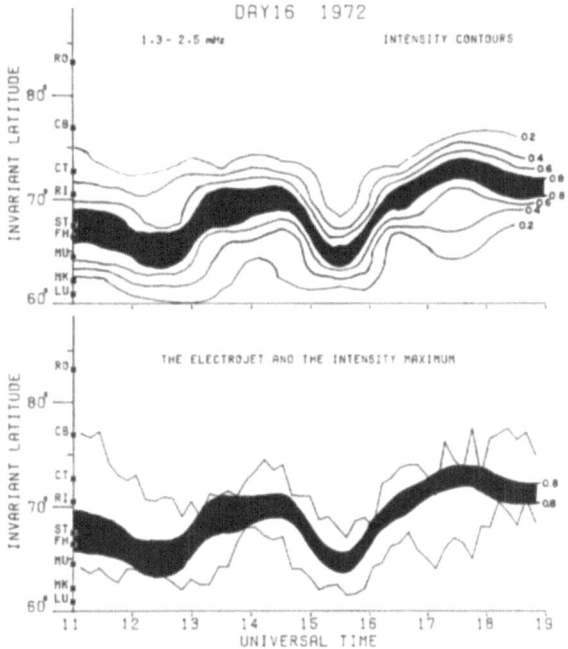

Fig. 8. Isointensity contours for Pc 5 activity (top panel) and the regime where the pulsation intensity is >80% of peak intensity together with the contours of the poleward and equatorward boundaries of the westward electrojet (after LAM and ROSTOKER, 1978). Note how the Pc 5 activity is confined within the electrojet boundaries.

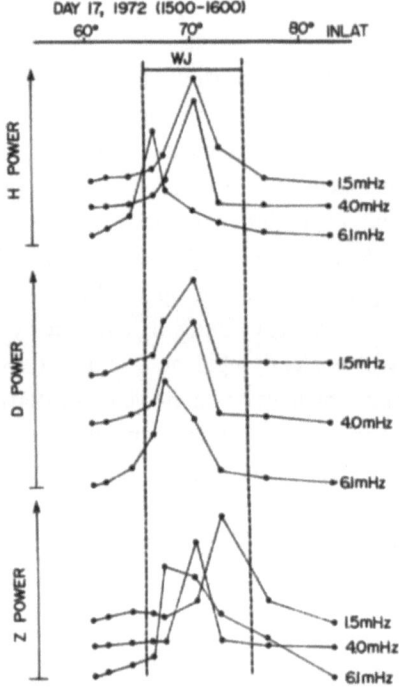

Fig. 9. Power in the three components of the pulsation field at three different frequencies during an hour long interval. The boundaries of the westward electrojet are indicated by the vertical dashed lines (after Lam and Rostoker, 1978). Note the suggestion of a frequency-latitude dependence in the Z-component.

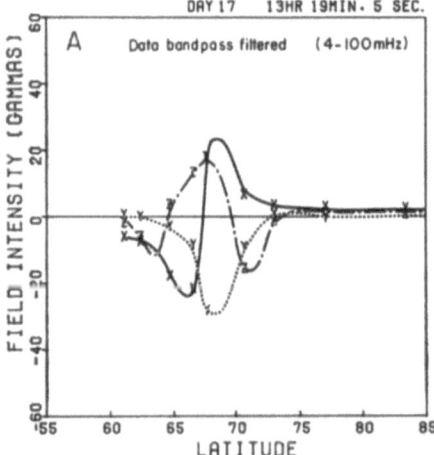
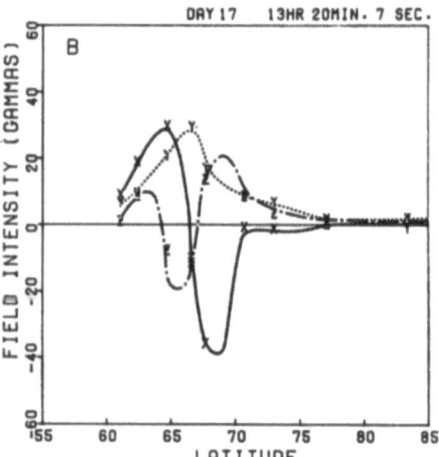

Fig. 10. Latitude profiles of the magnetic perturbation pattern associated with Pc 5 micropulsations at two instants one half cycle apart (after Rostoker and Lam, 1978).

variation of the magnetic perturbation pattern for Pc 5 pulsations at two instants one half cycle apart in time. In terms of equivalent currents, this pattern can be envisioned as being produced by antiparallel ionospheric currents, with the peak intensity of the pulsation located at the boundary between eastward and westward equivalent current flow.

ROSTOKER and LAM (1978) have attempted to explain the observed perturbation pattern in terms of oscillations of the auroral electrojet system associated with sudden changes in the level of magnetospheric convection.   The treatment of micropulsations as LC-oscillations of three dimensional current systems is a useful way in which to couch any model for the provision of energy for the pulsations.   In the context of this model, the energy for the pulsations is resident in the drifting magnetospheric plasma, with the drift energy of the plasma being gradually extracted to provide the electrical energy for the oscillating current system.   This leads to an explanation of Pc 4, 5 micropulsations in which the energy for the pulsations is resident on auroral oval field lines, and thus the energy source is inside the magnetosphere rather than being provided through a Kelvin-Helmholtz instability on the magnetopause.   It may well be that the observed Pc 4, 5 spectrum contains contributions from energy sources both internal to the magnetosphere and at the magnetopause.   Further careful research into the Pc 5 spectrum is necessary to clarify this issue.

4.   The Response of Dayside Pc 5 Activity to Nightside Substorms

One of the interesting questions about Pc 4, 5 activity centers around the regulation

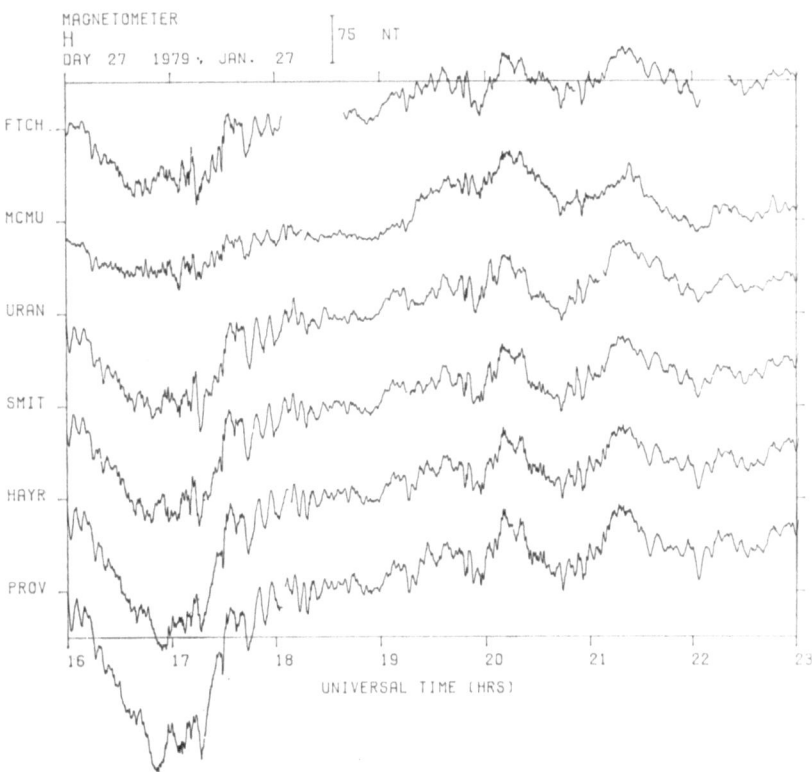

Fig. 11.   *H*-component magnetograms from the University of Alberta IMS magnetometer array for Day 27, 1979 (after SAMSON and ROSTOKER, 1980).   Note the irregular appearance of the Pc 5 activity across the local noon sector ~1900 UT).

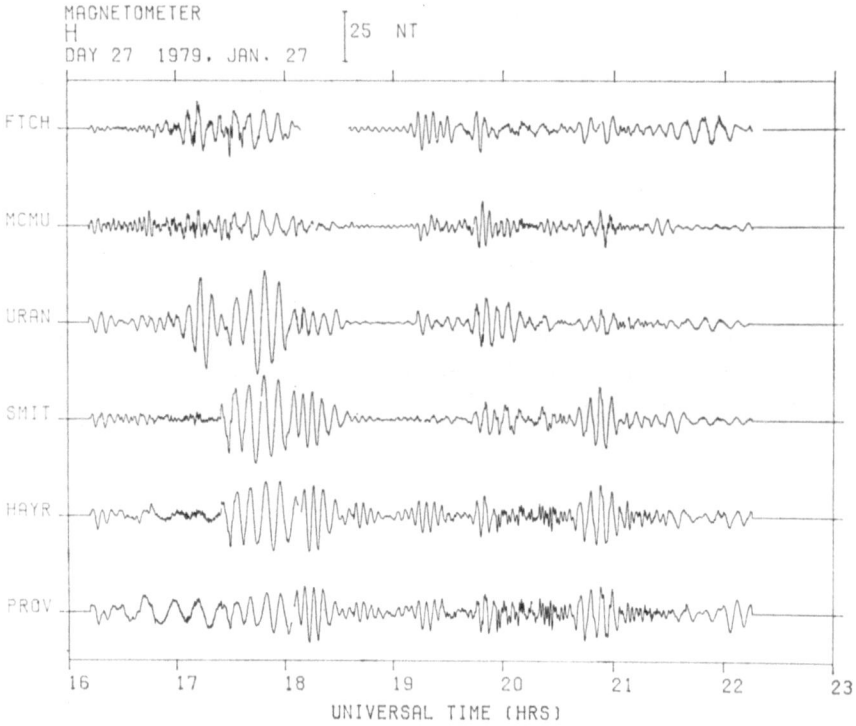

Fig. 12. Pc micropulsation activity in the frequency range $f > 1$ mHz corresponding to the
data shown in Fig. 11 (after Samson and Rostoker, 1980). Note the pronounced
changes in frequency content at ~ 1805, ~ 1835, ~ 2015, and ~ 2104 UT.

of the level of the power in the various portions of the pulsation frequency spectrum.
The impulsive nature of the Pc 5 activity discussed earlier suggests either impulsive changes
in the energy supply or in the level of coupling to the field lines on which the resonance
takes place. Lam and Rostoker (1978) have shown that the Pc 5 pulsation activity
appears to be enhanced when there is a decline in the level of substorm activity in the
magnetosphere as reflected by a decrease in the strength of the westward electrojet in the
pre-noon quadrant. A similar behaviour was reported for giant Pc 4 pulsations by
Rostoker et al. (1979), who also noted that these large amplitude pulsations were suppressed in conjunction with the onset of substorm activity in the nightside magnetosphere.
The possibility that substorm activity can affect dayside Pc 4, 5 activity had led to a study
of the response of Pc 4, 5 activity in the noon sector to substorm onsets in the midnight
sector carried out by Samson and Rostoker (1980). They found that, within a time scale
of as little as two to three minutes, a substorm onset near midnight was accompanied by
an increase in the dominant Pc 5 frequency across the noon sector.   This behaviour is demonstrated in Figs. 11, 12, and 13.   Figure 11 shows the H-component magnetograms
from the Alberta IMS array for the hours across local noon (~ 1900 UT in the Alberta
sector).   There is clearly considerable amount of Pc 5 pulsation activity, but from these
normal magnetograms very little trend is evident.   Figure 12 shows the same data to

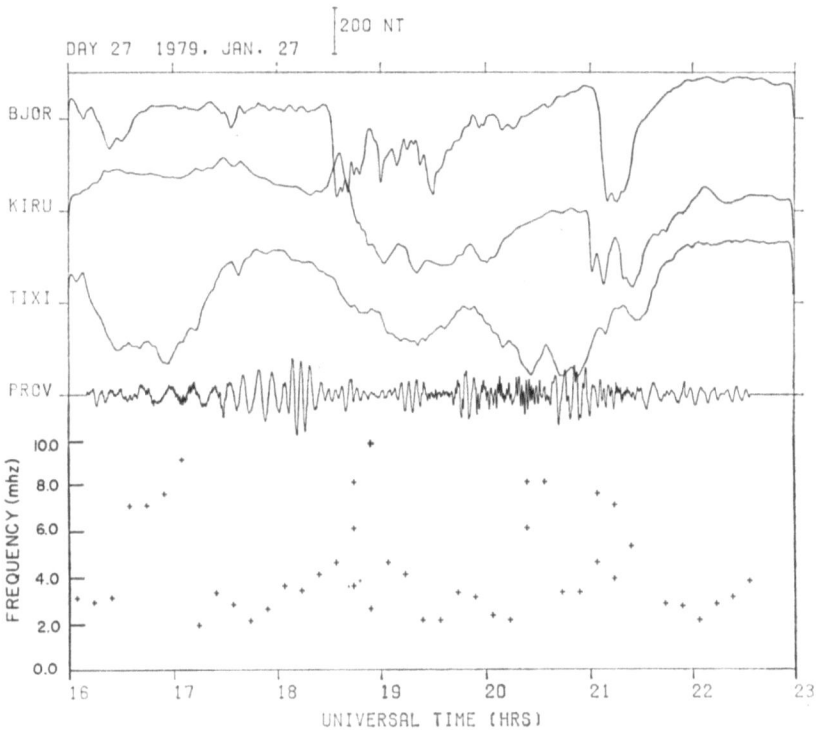

Fig. 13. *H*-component magnetograms from the Scandinavian observatories of Kiruna and Bjornøya and the Soviet station of Tixie Bay, together with the pulsation data for the noon sector auroral oval (Fort Providence) and maximum entropy estimates of the dominant Pc frequency (after SAMSON and ROSTOKER, 1980).

which a data adaptive polarization filter (SAMSON and OLSON, 1980) has been applied. This filter selectively removes the random noise in the magnetometer records leaving only the purely polarized signals in the frequency range of interest. In these data, very clear changes in frequency are evident from time to time, which leads one to ask if the frequency changes are correlated with any other magnetospheric phenomena. In fact, SAMSON and ROSTOKER (1980) have clearly shown that the dominant frequency in the Pc spectrum increases significantly in association with a substorm onset in the midnight sector. An example of this correlation is shown in Fig. 13 where several sample nightside *H*-component magnetograms, the pulsation data from PROV (obtained using the data adaptive polarization filter) and the variation of the dominant pulsation frequency (obtained using maximum entropy techniques) are shown. Key times on these magnetic data are noted to be ∼1805, ∼1835, ∼2015, and ∼2104 UT, when there were sharp increases in the dominant pulsation frequency. Normal magnetogram data from the Scandinavian sector indicate that clear substorm onsets or intensifications took place at ∼1805, ∼1835, and ∼2104 UT. The change in the pulsation frequency takes place close to local noon within less than 5 minutes of a substorm onset near local midnight. If the pulsation response is indeed triggered by the substorm near midnight, the information must travel through the magnetosphere-ionosphere system at approximately the Alfvén velocity. Finally we note the 2015 UT event, in which the marked increase in pulsation frequency appears to coincide with a decline in activity as measured by the auroral zone magnetogram, recorded in the Scandinavian sector (Kiruna and Bjornøya). In this case, it was found that the change in pulsation frequency coincided with a sudden equatorward motion of the auroral oval which was visible in both the auroral data and from magnetometer data on both the dayside and the nightside. In fact the auroral zone magnetogram for the central USSR sector shows a marked intensification of the westward electrojet at this time. This is an excellent example of how a sudden change in the position of the auroral electrojet produces a perturbation in the average auroral zone which might incorrectly be interpreted as a recovery phase.

In conclusion, we note that LAM and ROSTOKER (1978) and ROSTOKER et al. (1979) have tended to identify the onset of Pc 5 activity with the beginning of a recovery phase from previous activity. From the data presented here, it is clear that recovery from active conditions leads to the appearance of low frequency Pc 5 activity while the onset of substorm activity leads to a suppression of the low frequency activity in favour of the higher frequency components of the spectrum. Since the high frequencies normally have lower amplitudes than their low frequency counterparts, eyeball inspection of magnetograms would lead one to believe that all Pc 5 pulsation activity is suppressed in association with increase in substorm activity even though it is clear from the spectral analysis that the effect is more of a frequency shift.

5. Conclusions

In this review we have tried to emphasize two important aspects of Pc 4, 5 pulsation activity. Firstly, we have seen the wavelike character of the pulsations both along magnetic fields and perpendicular to those field lines in the azimuthal directions. In terms of source mechanisms, we have noted that the phase characteristics are consistent with the Kelvin-Helmholtz instability as a source of the wave energy. However, we have also pointed out some characteristics of Pc 4, 5 activity (i.e. local time variation of polarization and occurrence frequency) which are not in agreement with the expected behaviour if the Kelvin-Helmholtz instability were, in fact, the only source of Pc 4, 5 energy. Secondly, we have shown that, for the most part, the Pc 4, 5 pulsations source regions are located inside the boundaries of the westward auroral electrojet in the morning sector. In addition, sudden changes in the magnetosphere-ionosphere current system responsible for substorm variations appear to trigger changes in the frequency spectrum of Pc 4, 5 activity. Thus there is some evidence that changes internal to the magnetosphere may be a causative agent for the Pc activity. Some considerable amount of research is yet to be done before it will be known if the source of Pc 4, 5 activity in the magnetosphere is internal or external (or whether both types of sources are operative).

This research was supported by the Natural Sciences and Engineering Research Council of Canada and in part by the Department of Energy, Mines and Resources (Earth Physics Branch).

REFERENCES

CHEN, L. and A. HASEGAWA, A theory of long-period magnetic pulsations, 1. Steady state excitation of field line resonance, *J. Geophys. Res.*, **79**, 1024–1032, 1974a.

CHEN, L. and A. HASEGAWA, A theory of long-period magnetic pulsations, 2. Impulse excitation of surface eigenmode, *J. Geophys. Res.*, **79**, 1033–1037, 1974b.

DUNGEY, J. W., Electrodynamics of the outer atmosphere, *Ionos. Res. Lab. Sci. Rept.*, **69**, Pennsylvania State Univ., 1954.

GREEN, C. A., The longitudinal phase variation of mid-latitude Pc 3–4 micropulsations, *Planet. Space Sci.*, **24**, 79–85, 1976.

GUPTA, J. C., Some characteristics of large amplitude Pc 5 pulsation, *Aust. J. Phys.*, **29**, 67–79, 1976.

HERRON, T. J., Phase characteristics of geomagnetic micropulsations, *J. Geophys. Res.*, **71**, 871–889, 1966.

HUGHES, W. J., R. L. McPHERRON, and J. N. BARFIELD, Geomagnetic pulsations observed simultaneously on three geostationary satellites, *J. Geophys. Res.*, **83**, 1109–1116, 1978.

JACOBS, J. A. and K. O. WESTPHAL, Geomagnetic micropulsations, in *Physics and Chemistry of the Earth*, Vol. 5, edited by L. H. Ahrens, F. Press, and S. K. Runcorn, 157 pp., The McMillan Company, New York, 1964.

JACOBS, J. A., Y. KATO, S. MATSUSHITA, and V. A. TROITSKAYA, Classification of geomagnetic micropulsations, *J. Geophys. Res.*, **69**, 180–181, 1964.

KATO, Y. and T. UTSUMI, Polarization of the long period geomagnetic pulsation Pc 5, *Rept. Ionos. Space Res. Japan*, **18**, 214–217, 1964.

KAWASAKI, K. and G. ROSTOKER, Perturbation magnetic fields and current systems associated with eastward drifting auroral structure, *J. Geophys. Res.*, **84**, 1464–1480, 1979.

LAM, H.-L. and G. ROSTOKER, The relationship of Pc 5 micropulsation activity in the morning sector to the auroral westward electrojet, *Plant. Space Sci.*, **26**, 473–492, 1978.

OLSON, J. V. and G. ROSTOKER, Longitudinal phase variation of Pc 4–5 micropulsations, *J. Geophys. Res.*, **83**, 2481–2488, 1978.

ROSTOKER, G. and H.-L. LAM, A generation mechanism for Pc 5 micropulsations in the morning sector, *Planet. Space Sci.*, **26**, 493–515, 1978.

ROSTOKER, G., H.-L. LAM, and J. V. OLSON, Pc 4 giant pulsations in the morning sector, *J. Geophys. Res.*, **84**, 5153–5166, 1979.

SAMSON, J. C., Three-dimensional polarization characteristics of high-latitude Pc 5 geomagnetic micropulsations, *J. Geophys. Res.*, **77**, 6145–6160, 1972.

SAMSON, J. C. and J. V. OLSON, Data adaptive polarization filters for multichannel geophysical data, *Geophysics*, submitted, 1980.

SAMSON, J. C. and G. ROSTOKER, Latitude-dependent characteristics of high-latitude Pc 4 and Pc 5 micropulsations, *J. Geophys. Res.*, **77**, 6133–6144, 1972.

SAMSON, J. C. and G. ROSTOKER, Response of dayside Pc 5 pulsations to substorm activity in the nighttime magnetosphere, *J. Geophys. Res.*, 1980 (in press).

SAMSON, J. C., J. A. JACOBS, and G. ROSTOKER, Latitude-dependent characteristics of long-period geomagnetic micropulsations, *J. Geophys. Res.*, **76**, 3675–3683, 1971.

SOUTHWOOD, D. J., Recent studies in micropulsation theory, *Space Sci. Rev.*, **16**, 413–425, 1974.

Observations of Pc Pulsations in the Magnetosphere: Satellite-Ground Correlation

Susumu Kokubun

Geophysics Research Laboratory, University of Tokyo, Tokyo, Japan

(Received June 28, 1980)

Observations of Pc waves in the magnetosphere are reviewed with emphasis on the discussion of satellite-ground correlations. Gross features of wave occurrence in the Pc 3–5 frequency range are shown to be well summarized by taking into account the polarization characteristics of the waves with respect to the ambient magnetic field. In the Pc 4–5 frequency range, azimuthally polarized-transverse waves, which occur predominantly on the morning side of the magnetosphere, show a good correlation with ground Pc events. Radially polarized waves dominate in the afternoon and dusk sectors and are less correlated with ground pulsations than azimuthal waves. Compressional waves, such as stormtime Pc 5 observed at synchronous altitude in the afternoon and compressional Pc 5 observed with HEOS 1 in the dusk, have not yet been identified on the ground. Only compressional waves in space, which belong to the radial class and show a strong ground correlation, are giant pulsations as observed in the morning sector on the ground. Although the problem remains to be studied further, the difference in degree of ground-satellite correlations of Pc waves seems to reflect the spatial extent of respective wave phenomena. As for most of ground-correlated Pc waves, observations appear to yield evidence that a rotation of the wave ellipse orientation occurs between the magnetosphere and the ground as predicted by theory.

1. Introduction

In the last decade, rapid progress has been achieved in the observations of ULF waves both in space and on the ground. In situ observations, especially at the earth-synchronous orbit, have revealed the existence of various modes of Pc waves (e.g. Cummings et al., 1969; Barfield et al., 1972; Bosson et al., 1976a, b; Arthur et al., 1977; Arthur and McPherron, 1977a, b; Hughes et al., 1978). Statistical studies have shown that most of Pc 3–5 waves are transverse waves with respect to the ambient magnetic field and that they can be classified into the two types, azimuthally and radially polarized waves, according to the orientation of the wave ellipse (Arthur et al., 1977; Arthur, 1978). Radially polarized Pc 4–5 waves with a significant compressional component have also found in the region from the late afternoon to midnight (Barfield and McPherron, 1972b; Hughes et al., 1979). Oscillations of energetic particle fluxes in the Pc 4–5 frequency range and their relation to magnetic field waves have also been examined recently by Su et al. (1977, 1979), Kokubun et al. (1977), Cummings et al. (1978), Higbie et al. (1978), and Hughes et al. (1979). Although the early earth-based conjugate point studies demonstrated that the low frequency Pc waves primarily show standing wave structures along the field line (e.g. Lanzerotti and Fukunishi, 1974), it is quite recent that simultaneous

17

measurements of magnetic field and particle flux variations in the magnetosphere yield convincing evidence of the standing wave structure of Pc 4–5 waves (KOKUBUN et al., 1977; CUMMINGS et al., 1978; SINGER and KIVELSON, 1979).

On the other hand, the use of arrays of closely spaced magnetometers on the ground in the early 70's revealed the spatial characteristics of Pc 4–5 pulsations (SAMSON et al., 1971) and made an important impact on the development of theories. CHEN and HASE-GAWA (1974) and SOUTHWOOD (1974) independently pointed out that the observed signal structure was consistent with a theory that combined the current ideas on the Kelvin-Helmholz instability source at the magnetopause with the concepts of field line resonance. The subject related to this resonance theory has extensively been reviewed by LANZEROTTI and SOUTHWOOD (1979).

Theoretical considerations have also showed that the ionospheric modification of magnetospheric signals should be taken into account to infer the nature of magnetospheric ULF waves from the ground observations (INOUE, 1973; HUGHES, 1974; HUGHES and SOUTHWOOD, 1976a, b). One of important results of these theories is that the ground observations are limited to detecting magnetospheric signals of a scale length less than ~ 120 km at the ionospheric level. In fact, some of magnetospheric ULF waves, such as stormtime Pc 5, have not yet been identified on the ground (BARFIELD and MCPHERRON, 1972b; HEDGECOCK, 1976). Although present ground stations are not sufficient for examining satellite-ground correlations of ULF waves, especially of short period waves, this may indicate that there exists a localized ULF wave in the magnetosphere which is not observed on the ground. Multi-satellite observations have indeed shown that certain modes of oscillations in space occur in a spatially limited region (HUGHES et al., 1979; SINGER et al., 1979). Since the opportunity to use simultaneous data from multi-satellites is, however, rather limited, it is important to establish the ground correlations of magnetospheric signals in order to make use of single point satellite measurements.

In this paper we will first discuss characteristics of Pc 3–5 waves, such as polarization and local time distributions of occurrence, based on the recent spacecraft observations. The satellite-ground correlations of these waves are also reviewed. It is shown that the classification of Pc waves by the direction of the wave ellipse with respect to the ambient magnetic field (ARTHUR et al., 1977) is very useful for summarizing the occurrence characteristics of Pc 3–5 waves in space and their satellite-ground correlations.

2. Polarization and Occurrence Characteristics of Pc Waves in the Magnetosphere

Characteristics of Pc waves in the magnetosphere are first summarized, since it is important to know about polarization properties of these waves in space for the following discussions on satellite-ground correlations. Statistical examinations of wave properties have mainly been made by using data from geostationary satellites (e.g. ARTHUR et al., 1977; BARFIELD and MCPHERRON, 1972b; BOSSON et al., 1976b). ARTHUR et al. (1977) have found that Pc 3 waves observed at ATS 6 can be classified into two types of waves with different characteristics, by taking into account the direction of the major axis of polarization. Azimuthal waves, which comprise the main portion of the dominant frequency observations, show a local-time frequency dependence (with higher frequencies occurring near noon), and are linear and very transverse. Radial waves, which mainly occur at frequencies higher than 0.03 Hz, show no variation of frequency with local time, and are linear but often have a significant compressional component. This classification

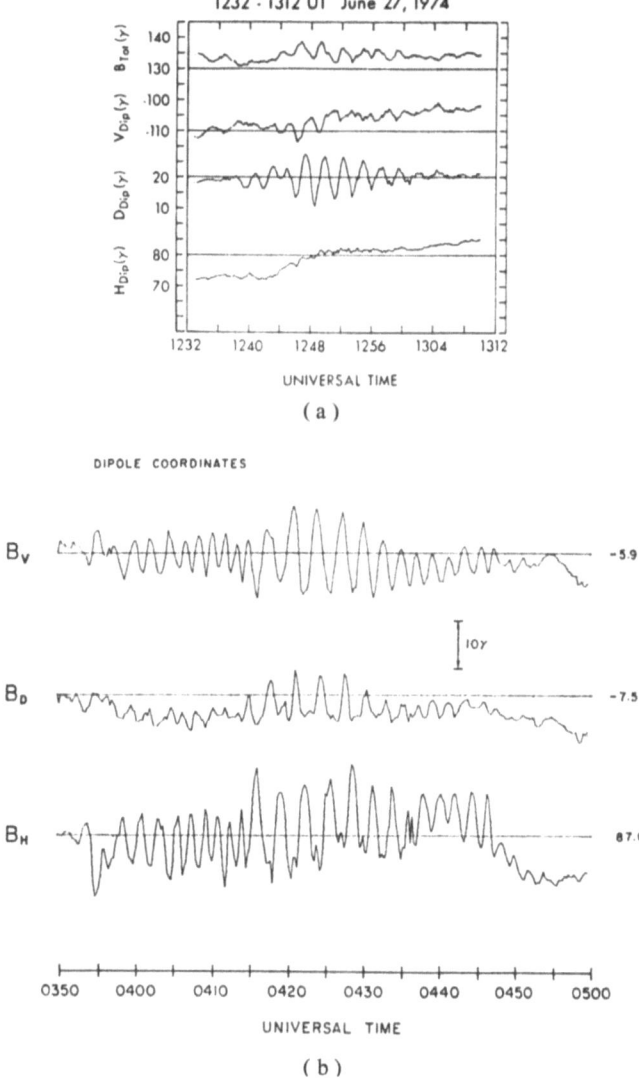

Fig. 1. Examples of time-amplitude records of azimuthally and radially polarized Pc waves near $L \simeq 6.6$. (a) A transverse and azimuthally polarized Pc 4 wave observed at ATS 6 (after CUMMINGS et al., 1978). (b) A stormtime Pc 5, which is a radially polarized with a significant compressional component, observed at ATS 1 (after BARFIELD et al., 1972). Both plots are made in the dipole VDH coordinate system. The H axis is antiparallel to the earth's dipole axis, D is azimuthally east perpendicular to the dipole meridian plane through the satellite, and $V = D \times H$ is roughly radially outward.

is very useful to characterize the occurrence of Pc waves in the other frequency range, as will be shown.

Figure 1 shows two examples of time-amplitude records of Pc 4 and 5 waves obtained at ATS 1 and 6 (BARFIELD *et al.*, 1972; CUMMINGS *et al.*, 1978). Plots are made in the dipole coordinate system, in which the H axis is antiparallel to the earth's dipole axis. D is azimuthally east, perpendicular to the dipole meridian plane through the satellite, and V is perpendicular to both H and D, and is roughly radially outward. The Pc 4 wave shown in the upper panel can be categorized as an azimuthal wave (A-class), because the amplitude dominates in D, and the compressional component is small. The measurement of the Poynting vector of this wave, using simaltaneous modulations of low-energy protons, gives convincing evidence that the wave is a standing hydromagnetic wave along the magnetic field (CUMMINGS *et al.*, 1978). A-class Pc 5 waves with characteristics similar to this wave were identified mostly in the morning sector of $L=6\sim12$ in the ULF wave surveys of OGO 5 data by KOKUBUN *et al.* (1976, 1977) and SINGER and KIVELSON (1979). Also for selected events the magnetic field observations have been correlated with other measurements such as energetic particle flux and cold plasma density, and wave electric fields have been inferred. They have also concluded that A-class Pc 5 waves are standing Alfvén waves and that the dominant mode of oscillation is the fundamental odd mode.

The stormtime Pc 5 illustrated in the lower panel shows different characteristics from that of A-class waves. Amplitudes are larger in H and V, and the wave is nearly confined to the magnetic meridian plane. Waves of this type may be categorized as R-class waves. Stormtime Pc 5 waves are a class of hydromagnetic waves observed at synchronous orbit most frequently in the afternoon sector during the main phase of magnetic storms and usually accompany a depression in local field magnitude associated with the onset of a magnetospheric substorm. (BARFIELD and MCPHERRON, 1972b, 1978). It was also found that Pc 1 activity of large amplitude is often superimposed on the stormtime Pc 5 (BARFIELD and MCPHERRON, 1972a). HEDGECOCK (1976) reported in his analysis of HEOS-1 magnetic field data that compressional waves, which have polarization characteristics similar to those of stormtime Pc 5, were observed in the dusk sector of L values between 8 and 12. Although a depression of the ambient magnetic field always accompanies these compressional Pc 5 events, they are not obviously related to periods of enhanced magnetic activity.

ARTHUR (1978) has recently showed that Pc 4 waves, observed at ATS 6, can be divided into 4 classes on the basis of frequency and azimuth of the major axis of polarization and that these 4 classes have quite different local time distributions. The waves of A-class in the higher frequencyrange of the Pc 4 band ($0.015<f<0.022$ Hz) are mostly observed in the daytime around noon. Although she reported that R-class waves are classified into two types according to the frequency, both classes predominantly occur on the dusk side around 1800 LT. Lower frequency Pc 4 waves ($f\leq0.015$ Hz) with an azimuthal polarization occur mostly around 0600 LT in the morning.

In Fig. 2 are schematically summarized local time distributions of Pc wave occurrence in the region of L values larger than 6, including occurrence characteristics of Pc 1 waves (BOSSON *et al.*, 1976a, b; KAYE and KIVELSON, 1979). The occurrence characteristics of A-class waves are expressed by thick lines, and those of R-class waves by dotted lines.

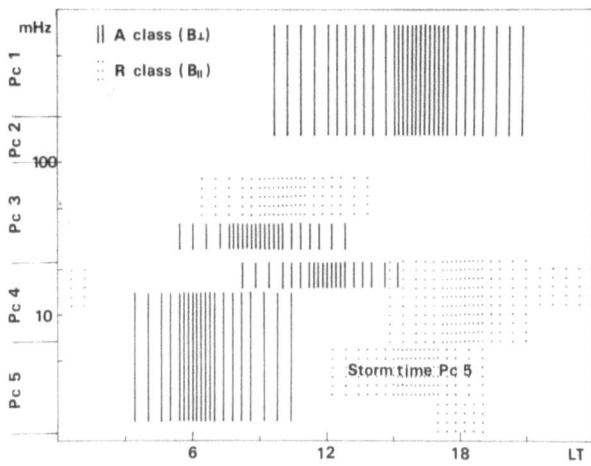

Fig. 2. Local time distributions of Pc wave occurrence in the magnetosphere for L values larger than ~ 6. The occurrence characteristics of azimuthal class waves are expressed by thick lines and those of radial class waves by dotted lines.

Since R-class waves tend to have a significant compressional component, as compared with A-class waves, compressional waves are included into the R-class in this figure. It is clearly seen that gross characteristics of Pc waves occurrences in space are well summarized by taking into account the classification scheme of ARTHUR et al. (1977). A-class waves in the Pc 3–5 frequency range tend to dominate in the morning and daytime sectors, and R-class waves in the Pc 4 and 5 range are mostly observed in the afternoon and the dusk. It is important to note that the azimuthal wave number of an A-class wave in the morning is smaller than that of a R-class wave in the afternoon, as shown by multi-satellite observations (HUGHES et al., 1978).

It should also be worth mentioning here that the latitudinal structure of hydromagnetic waves in space must be considered in interpretation of statistical characteristics such as the local time dependence of occurrence and the dominant period, especially for observations at the geostationary orbit. If Pc waves are the fundamental mode of standing oscillations along the field line, the transverse magnetic component should not be observed on the equatorial plane. Such a feature was first noted in their survey of Pc 5 waves by using OGO 5 magnetometer data (KOKUBUN et al., 1976, 1977). They found that transverse Pc 5 waves were not detected near the magnetic equator (latitude less than about 10°). SINGER and KIVELSON (1979) have further examined the latitudinal structure of Pc 5 waves by using simultaneous magnetic field and ion flux data.

Figure 3 shows two examples of their observations of Pc 5 waves at high and low magnetic latitudes. The top segments of the figure show the hydrogen ion flux on two different passes of OGO 5 through the plasma within a few hours of local dawn. In both cases, there were large-amplitude fluctuations in the ion flux with a period of 330 sec. Also shown is the magnetic component that is expected to be in quadrature with the flux modulation if the flux modulation is due to the presence of a standing Alfvén wave. The field oscillations in the left panel are $\sim 90°$ out of phase with the ions, indicating the

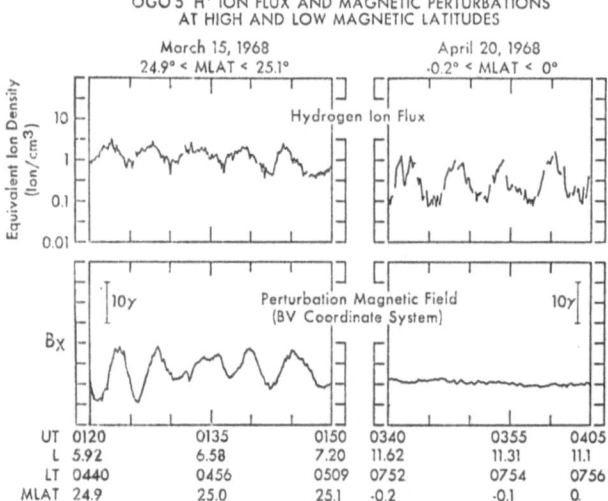

Fig. 3.   Two Pc 5 events as observed in ion flux oscillations.   The March 15 event at ∼25°
magnetic latitude has associated magnetic field oscillations, whereas no discernible
magnetic oscillation is associated with the April 20 event observed near the magnetic
equator (after SINGER and KIVELSON, 1979).

presence of a standing wave, which is most likely a fundamental field line oscillation since
the magnetic field perturbation peaks are on the falling slope of the ion fluctuation (KOKU-
BUN *et al.*, 1977).   The April 20 case in Fig. 3 might be interpreted as true density fluctuations because there was no identifiable magnetic wave activity related to the ion flux.
However, another possible explanation is that the satellite was near the magnetic mode
of an odd mode standing wave and that the ions were responding to the wave electric
field perturbation.   As will be shown in the next section, this hypothesis is supported
by the fact that ion fluctuations such as in the April 20 case shows a good correlation
with ground Pc 5 events.   Thus surveys near the geomagnetic equator that rely only on
magnetic field observations may miss many standing wave events (SINGER and KIVELSON,
1979).

Observations at sychronous altitude are ideal to examine statistical characteristics of
ULF waves, such as the dominant period near $L \simeq 6.6$ and the local time distribution of
wave occurrence, since synchronous satellites are fixed relative to the earth.   However,
the location of a satellite relative to the magnetic equator should be considered to interpret
statistical results, because the observation close to the magnetic equator may miss many
standing oscillation events.   In fact, statistical results of Pc 4 occurrence, based on
ATS 1 and 6, which were on the magnetic equatorial plane (CUMMINGS *et al.*, 1975;
HILLEBRAND and MCPHERRON, 1980), are considerably different from those obtained
by ARTHUR (1978), who used data from ATS 6 which was then at latitude of ∼10°.   Observations on the magnetic equator indicated that the maximum occurrence of Pc 4 is shortly
after local noon, and that the distribution of periods has a distinct maximum around
60–70 sec (CUMMINGS *et al.*, 1975; HILLEBRAND and MCPHERRON, 1980).   Although

a distinct peak in a shorter period range of the Pc 4 band, was also found in the analysis by ARTHUR (1978), other peaks were observed around periods of ~100 and 150 sec. These differences probably reflect the latitudinal structure of Pc waves.

3. Satellite-Ground Correlations of Pc 4–5 Waves

In the earlier studies, PATEL and CAHILL (1964) and PATEL (1965) compared transverse Pc 4–5 waves, observed at $L=6$–10, with ground observations near the subsatellite point, and found that correlations existed only when the satellite was near the same L shell and within one hour. LANZEROTTI and TARTAGLIA (1972) showed that a purely compressional wave, as observed at ATS 1 by BARFIELD et al. (1971), was detected near its conjugate point as an elliptically polarized transverse wave, although the power on the ground was more than 50 times smaller.

With respect to stormtime Pc 5 events observed at ATS 1, BARFIELD et al. (1972) concluded that these waves were not observed on the ground at the same time as their satellite observations at geostationary orbit on the magnetic equator. On the other hand, LANZEROTTI et al. (1974, 1975) reported that a ground-correlated Pc 5 event was identified at $L \simeq 5.5$ by Explorer 45, which was near 15° magnetic latitude and 2000 hours magnetic local time. They called this event a stormtime Pc 5, but it should be noted that the polarization characteristics of this wave differ considerably from those identified as a stormtime Pc 5 at ATS 1 by BARFIELD and McPHERRON (1972b) and BARFIELD et al. (1972). Strormtime Pc 5 waves at ATS 1 are R-class waves of which the major axis of the polarization ellipse is apparently confined to the magnetic meridian plane, while the wave observed at Explorer 45 is a A-class wave without significant compressional component, of which the major axis is predominantly in the azimuthal direction. An additional example of a ground-correlated Pc 4–5 event, observed with Explorer 45 in the afternoon sector, was reported by LIN and CAHILL (1976). They analysed wave event associated with a large scale magnetospheric compression during the recovery of a magnetic storm. A 2-mHz component was principally compressional at the satellite and was strongest on the ground at high-latitude stations. A 6-mHz component with the right-handed polarization (clockwise looking along the field line) and the major axis nearly in the azimuthal direction was also observed at auroral zone stations. However, pulsations near 15 mHz, which developed after the lower frequency waves at the satellite were not identified on the ground.

Although the number of data samples is not so large in the above studies, we note an interesting tendency that waves of the A-class are more correlated with ground events than waves of the R-class. In fact, compressional waves observed with HEOS-1, which have a polarization characteristic similar to stormtime Pc 5, were not identified on the ground (HEDGECOCK, 1976). Furthermore, the survey of Pc 5 waves using data from OGO 5 revealed that ground-correlated Pc 5 waves in the outer magnetosphere ($L=6 \sim 12$) belong to the A-class (KOKUBUN et al., 1976, 1977).

KOKUBUN et al. (1976) first compared wave-like variations at OGO 5 with ground magnetograms from high-latitude station, considering the possibility that low frequency fluctuations observed with an eccentric orbiter like OGO 5 are spatial rather than temporal.

From this comparison they found satellite-ground correlated events on 13 orbits and also identified Pc 5 waves, which showed characteristics similar to those of ground-correlated events, on 5 other orbits.   Figure 4 represents their unique example of Pc 5 wave events observed at the near-conjugate pair stations, Great Whale River ($L \simeq 6.3$) and Byrd ($L \simeq 9$), and simultaneously at OGO 5.   Two waves with periods of 255 and 215 sec are apparent in the intervals 1040–1110 and 1110–1135.   Perturbations are predominantly in B_y (east-west component) in the magnetosphere, while variations in the H component are larger than those in the D component on the ground.   Peak-to-peak amplitudes are approximately 9 nT at OGO 5 and 30–50 nT at the ground stations.   At OGO 5 the magnitude of compressional variations is less than one third of the transverse components. From the polarization analysis of approximately 40 wave packets they showed that

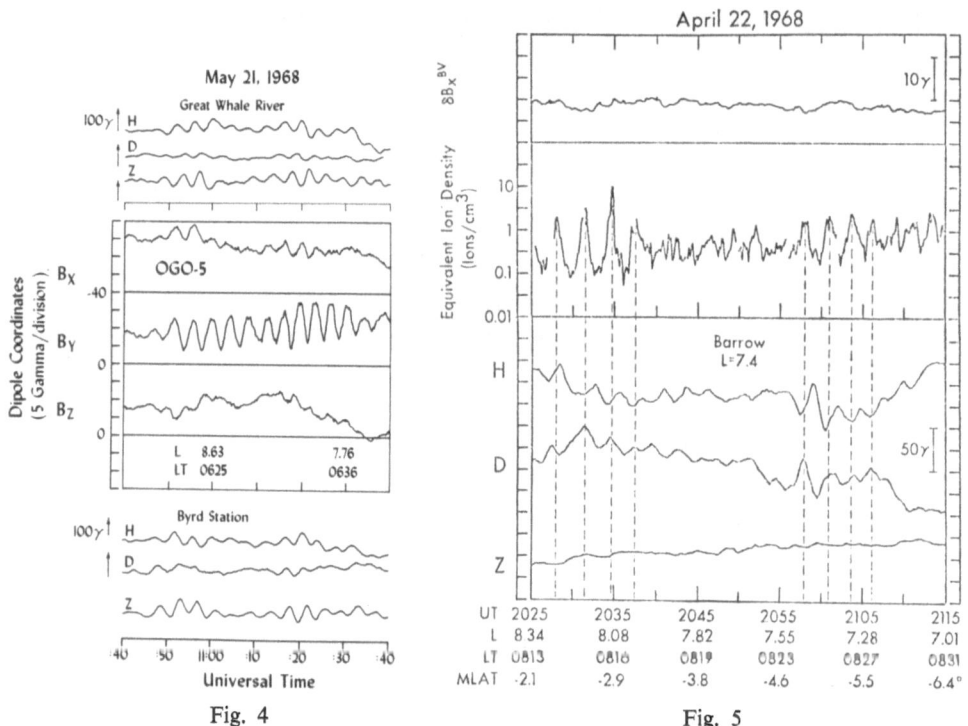

Fig. 4

Fig. 5

Fig. 4.   An example of ground-satellite correlations of Pc 5 waves as observed at OGO 5. OGO 5 was inbound at a magnetic latitude of $10° \sim 13°$ at 0620–0646 LT.   The conjugate pair stations, Great Whale River ($L \simeq 6.3$) and Byrd ($L \simeq 9$) were situated in the region of 0520–0630 LT.   The scale value of the Z component at Byrd is not available (after KOKUBUN et al., 1976).

Fig. 5.   Pc 5 ion flux oscillations observed by OGO 5 are shown in the middle panel.   Also displayed are simultaneous magnetic field records from (top) OGO 5 and (bottom) Barrow, Alaska, at the foot of the OGO 5 field line.   No discernible magnetic field oscillation at the period of the ion oscillations is observed by OGO 5, which is near the geomagnetic equator; however, large-amplitude magnetic oscillations are observed at the foot of the OGO 5 field line (after SINGER and KIVELSON, 1979).

Pc 5 waves identified in off-equatorial regions of latitudes $10°$ to $\sim 40°$ belong to waves of the A-class. The waves were mostly observed in the morning sector. The sense of polarization was mostly left-handed in the morning and right-handed in the afternoon, as expected from the ground observation (e.g. KOKUBUN and NAGATA, 1965; SAMSON, 1972).

Figure 5 shows another interesting observation obtained in the analysis of OGO 5 data by SINGER and KIVELSON (1979). In this figure are given simultaneous measurements of the ion flux at OGO 5 and of the magnetic field at Barrow, Alaska ($L=7.4$) near the foot of the OGO 5 field line. For each peak in the ion flux, there was a corresponding peak in either the H or D component at Barrow. There was no clear magnetic field activity at OGO 5 at the period of the ion oscillation, however, the amplitude on the ground became as large as ~ 50 nT peak to peak. These ground-satellite correlations are consistent with the hypothesis that the ion fluctuations were produced by a field line oscillation and that the satellite was near a magnetic node of the wave.

These OGO 5 observations have greatly helped to improve our understandings of the characteristics of Pc 5 waves, such as the latitudinal structure, and correlations with ground observations. However, the sample size was not large enough for satisfactory statistics especially concerning polarization. In addition there is some inconsistency in investigations using data from eccentric orbiters. HEDGECOCK (1976) found compressional Pc 5 waves in the dusk sector, but little or no Pc 5 in the morning sector. Thus there is much work yet to be done in characterizing magnetospheric Pc 5 waves statistically.

Analyses have recently been made independently by HILLEBRAND and McPHERRON (1980) and the author, to examine characteristics of Pc 4–5 waves at ATS 6 and their ground correlations. ATS 6 was located at 96°W in geographic longitude from June, 1974 to May, 1975, and was shifted to 35°E in the end of June, 1975 (McPHERRON *et al.*, 1975). The foot point of field line through ATS 6 in the former period, when the satellite was at magnetic latitude of $\sim 10°$, lies at a point between Churchill (63.8, 325.0, in geomagnetic coordinate) and Great Whale River (66.5, 349.5). The author examined data from June to December, 1974, and magnetograms from Churchill, Great Whale River and Thompson (66.8, 321.4). HILLEBRAND and McPHERRON (1980) also compared ATS 6 data with ground observations, using a more complete data set obtained from the Scandinavian meridian chain experiment in June, August and September, 1975, when the satellite was on the magnetic equator. The results obtained by the author will first be discussed in the following.

From the visual inspection of time-amplitude plots of the 5 sec average magnetic field from ATS 6 (McPHERRON, 1976), 67 wave events in the Pc 4–5 band were found to be simultaneously detected at one or more of the ground stations. Figure 6 shows an example of simultaneous Pc 5 activity, observed in the morning sector on September 1, 1974. A distinct wave packet with peak-to-peak amplitude of ~ 15 nT was apparent in the interval 1005–1045. The wave was predominantly in the east-west component at ATS 6 while variations in Z and D were larger than that in H on the ground. This ground signature differs considerably from that shown in the previous example (Fig. 4). However, such a feature is expected in case that a ground observatory is located close to

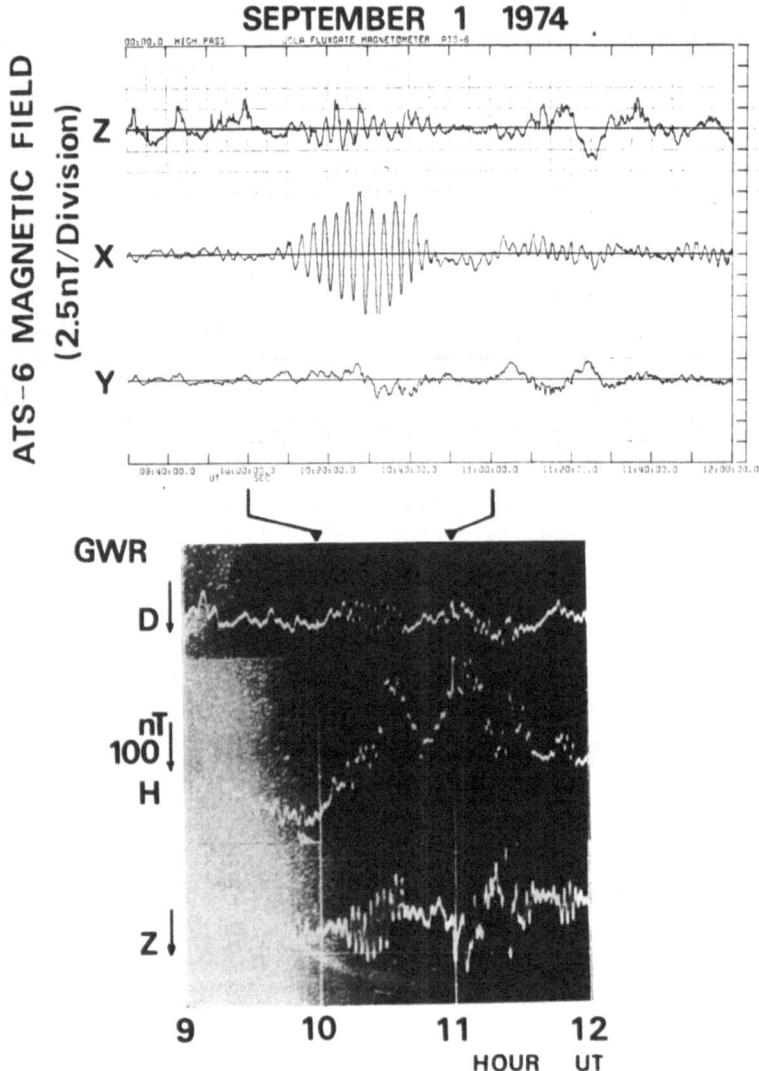

Fig. 6. Simultaneous Pc 5 activity observed at ATS 6 and Great Whale River on September
1, 1974. ATS 6 data are plotted in the satellite body coordinates. Z is roughly in the
radial direction, X in the east direction, and Y nearly antiparallel to the earth's dipole.
A distinct wave packet with a period of ~ 170 sec is seen in X in the interval 1005–
1045 UT (0355∼0435 LT).

the center of a localized signal and that a signal varies rapidly in the direction of east-west,
as theoretically discussed by SOUTHWOOD and HUGHES (1978).   Generally, ground Pc 4–5
signals, which correlate with waves in space, have the largest amplitude in the H component, as shown in Fig. 7.   Waves at ATS 6 were large in the azimuthal component,
as similar to previous examples, while variations at Thompson showed the usual characteristics of Pc 4–5 at high latitude.

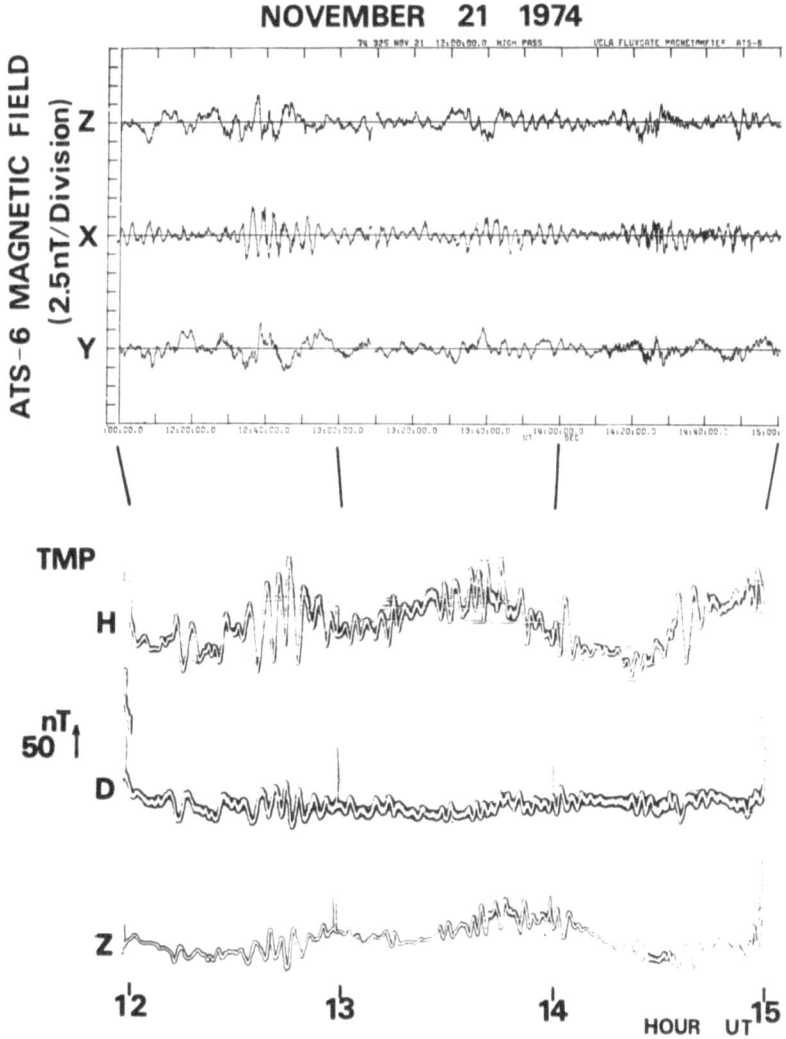

Fig. 7. Simultaneous records of Pc 5 waves at ATS 6 and Thompson. The wave amplitude
at ATS 6 dominates in the azimuthal component, as similar to that shown in Fig. 6,
while ground wave has a large amplitude in the *H* component.

As will be shown most of the ground-correlated Pc 4–5 waves at synchronous altitude
occur in the morning sector.   However, isolated ground-correlated signals are sometimes
observed in the afternoon.   Figure 8 represents one of such wave events.   As is similar
to those observed in the morning, the wave has its largest amplitude (~ 7.5 nT) in the
azimuthal component in the magnetosphere.   On the ground amplitudes in *H* and *D*
components are larger than 50 nT.   Although the detailed polarization analysis has not
yet been made, almost all of ground-correlated waves at ATS 6 were found to be an
azimuthally polarized wave from the visual inspection of 5 sec data plots.   We could
not identify ground signals corresponding to the wave of R-class, such as stormtime Pc 5,

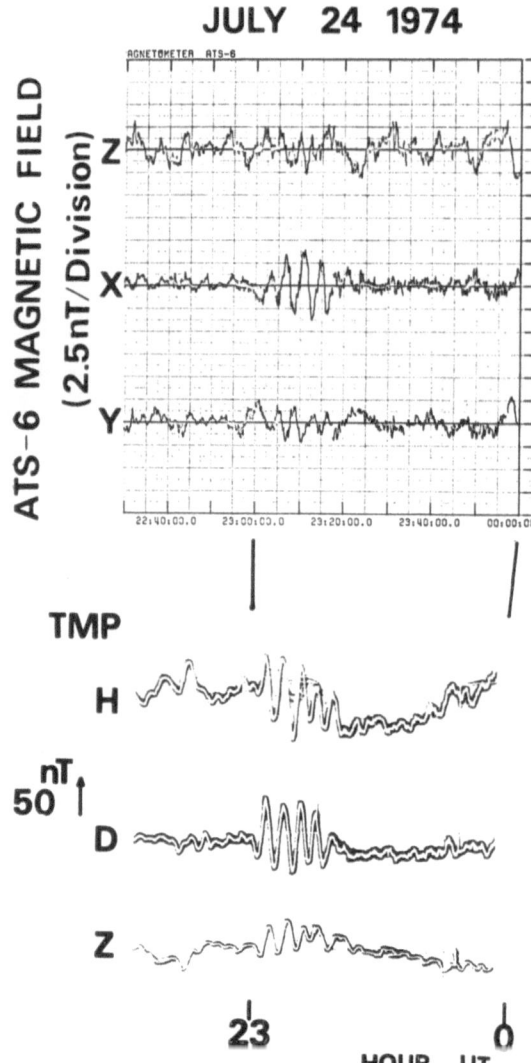

Fig. 8. Example of an isolated Pc 5 wave observed in the afternoon. The wave in space has a larger amplitude in the azimuthal component, similar to waves observed in the morning (see Figs. 6 and 7).

within resolution of ordinary magnetograms.

The local time distribution of ground-correlated waves at ATS 6 and the K_p dependence of their occurrence are shown in Fig. 9. The occurrence has a maximum around 0700 LT and a small peak around 1600. At the geostationary orbit the waves tend to appear in moderately disturbed conditions, whereas Pc 5 waves at OGO 5, reported by KOKUBUN *et al.* (1976) and SINGER and KIVELSON (1979), were observed mostly in the more quiet condition of $K_p \simeq 2$. This suggests that the center of Pc 5 oscillation shifts inward associated with magnetic activity.

Figure 10 shows the period distribution of ground-correlated Pc 4–5 waves at ATS 6. The periods of the morning waves mostly range in 140–170 sec, while those of afternoon waves are scattered. In Fig. 11 is shown the distribution of periods of oscillations versus local time for 78 Pc 4 events observed at ATS 6 (HILLEBRAND and MCPHERRON, 1980). This distribution is considerably different from those for ground-correlated waves, shown in Figs. 9 and 10, although Pc 5 waves with periods larger than 200 sec were excluded in their analysis. In Fig. 11 the wave with periods around 150 sec is not noted in the morning sector. The dominant period is in the range 50–70 sec and the peak occurrence is in the early afternoon. It should be considered to interpret this difference that the result of Hillebrand and McPherron was based on the equatorial observations.

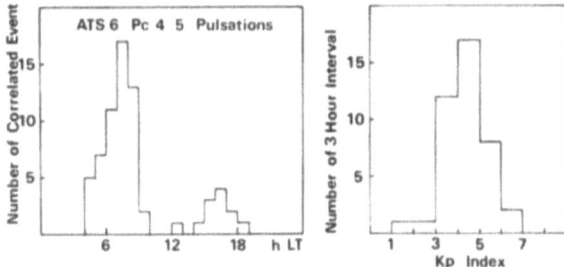

Fig. 9. Local time distribution and K_p dependence of occurrence of ground-correlated Pc 4–5 waves at ATS 6.

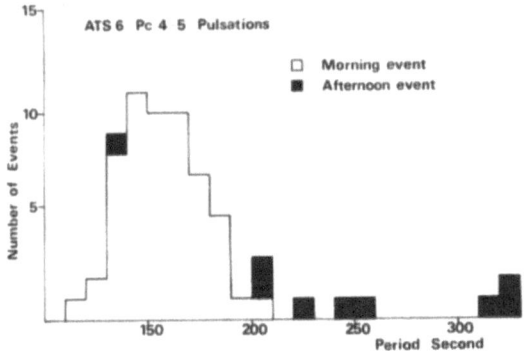

Fig. 10. Period distribution of ground-correlated Pc 4–5 waves observed at ATS 6.

From the coherency analysis they showed that 74% of the 78 Pc 4 waves were transverse but radially polarized. 10% were azimuthally polarized and 15% were compressional. The subdivision of the events into local time groups did not reveal a striking difference except for a lack of compressional events around noon.

As for satellite-ground correlations, they reported that approximately 53% of the 78 events in space were detected on the ground, and that the compressional and azimuthal classes had the largest relative number of ground-correlated events. Namely, 7 out of 8 azimuthal waves and 10 out of 12 compressional waves were noted as line spectra on the ground, while the ground detection was 38% for the transverse-radial waves. For the characteristics of power spectra they noted the following facts: The enhancements in the spectra at ATS 6 were usually large enough to enable a clear decision as to whether an event could be selected or not. However, most of the ground data had the prop-

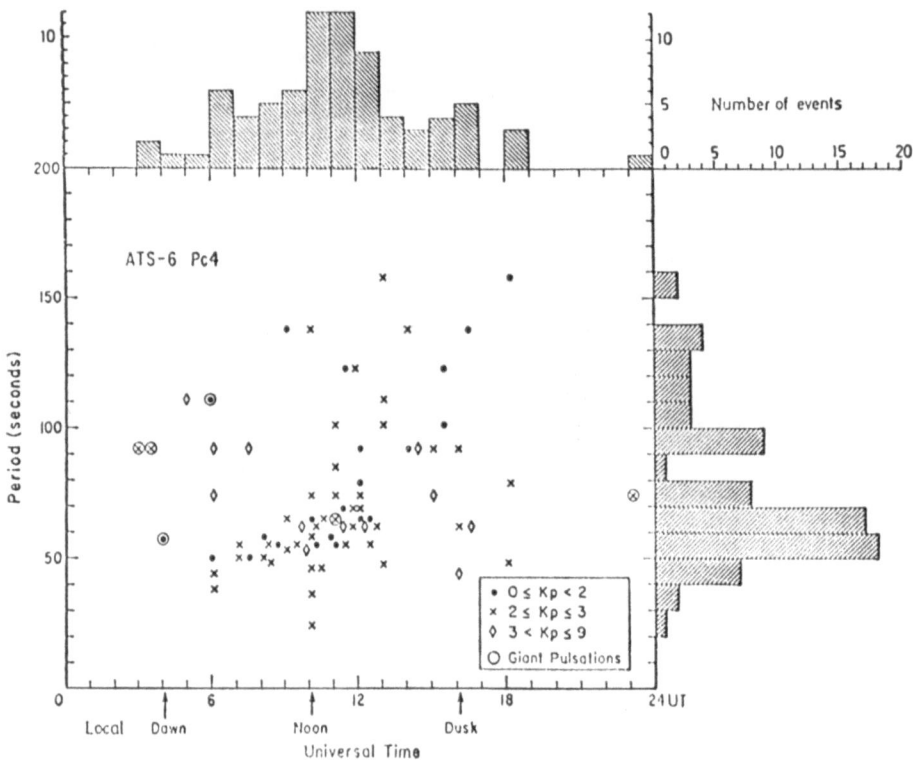

Fig. 11.   Period distribution and local time variation of occurrence for Pc 4 events observed
at ATS 6 at the geomagnetic equator (after Hillebrand and McPherron, 1979).

erties of noise spectra.   Few ground events have so clear spectra as giant pulsations, discussed in the next section.

The distribution shown in Fig. 11 is very similar to the result for transverse waves observed at synchronous-equatorial altitude (ATS 1) by Cummings et al. (1975).   They reported that the most frequently occurring period is in the range 60–70 sec and that the peak of the local time distribution is 1–3 hours after noon.   Cummings et al. (1975) have also argued that these transverse waves comprise the fundamental mode of a standing Alfvén wave near $L=6.6$, taking into account the measurement of the particle density by the detector aboard ATS 5 (DeForest and McIlwain, 1971).   However, further study is needed to give a basis on this argument, since the observation at off-equatorial latitude of the geostationary orbit show the distributions different from those reported by Cummings et al. (1975) and Hillebrand and McPherron (1980).   One of important differences is that azimuthally polarized-transverse waves with periods of 140–160 sec are rarely detected at equatorial-synchronous altitude.   Furthermore, convincing evidence has been obtained by Singer and Kivelson (1979) that A-class waves with the same characteristics as that observed at off-equatorial-synchronous altitude comprise the fundamental mode of a standing oscillation along the field line.   Therefore, the argument that transverse and probably radially polarized waves with the dominant period of 60–70 sec at ATS 1 are the fundamental mode of a standing Alfvén wave does not appear to be conclusive at present.

4.   Wave Characteristics of Giant Pulsations at Synchronous Altitude

In the Pc 4 band there exist a particular type of pulsations, known as giant pulsations (Pg), although their occurrence is relatively infrequent as compared with other types of pulsations. The most striking features of Pgs are their very sinusoidal appearance and the unusually long duration of the phase coherence-wave packets (see GREEN (1979) and ROSTOKER et al. (1979) for detailed characteristics and references). GREEN (1979) and ROSTOKER et al. (1979) have found in their analyses of the IMS magnetometer array data that these pulsations are highly localized in latitude and that have a large azimuthal wave number. The large magnitude of the azimuthal wave number does not necessarily reflect the longitudinal extent of the disturbed region (ROSTOKER et al., 1979). In this section are discussed recent results of characteristics of Pgs in space, independently

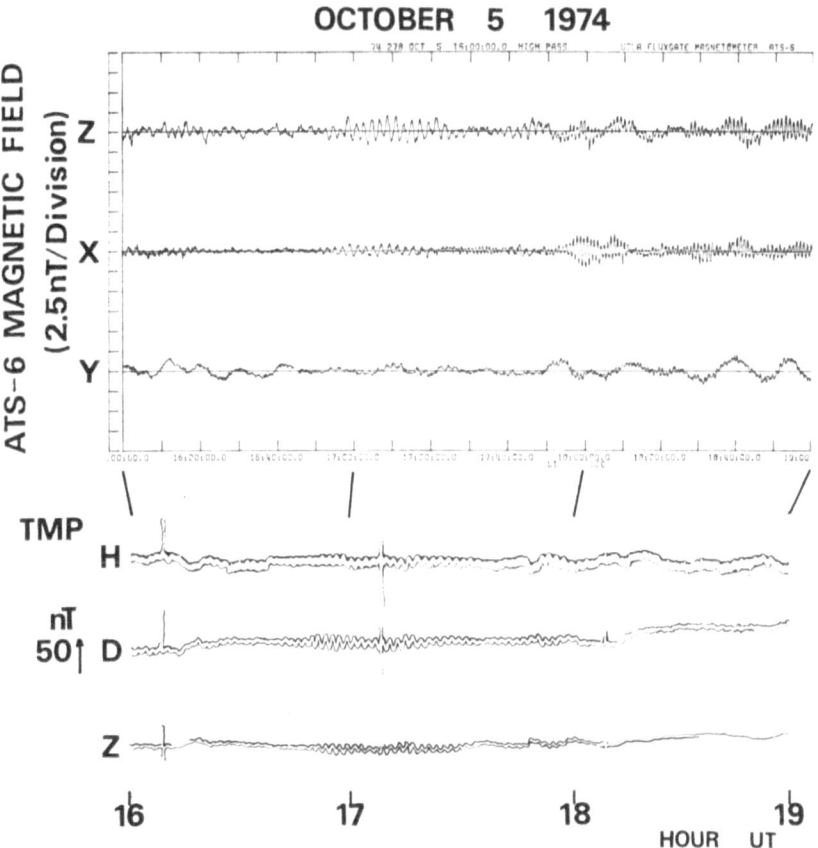

Fig. 12.   Simultaneous records of giant pulsations at ATS 6 and Thompson. At ATS 6 the amplitude of the Pg wave in the interval 1650–1730 is large in the radial component. The wave on the ground dominates in the D component.

obtained by HILLEBRAND and McPHERRON (1979) and by the author.

In his paper GREEN (1979) has inferred that the compressional wave event with a mean period of 106 sec, recorded on ATS 1 at synchronous altitude (BARFIELD *et al.*, 1971), represents an equatorial magnetospheric observation of a Pg near its resonance maximum. This suggestion was based on the analysis by LANZEROTTI and TARTAGLIA (1972), who found periodic fluctuations with the same period as the equatorial compressional wave at College about one hour in local time west of the foot of the magnetic meridian through ATS 1. Green have found that the polarization characteristics of this wave event are very similar to those of Pgs obtained in his analysis.

Observations at ATS 6 give support for Green's inference that a Pg is a compressional

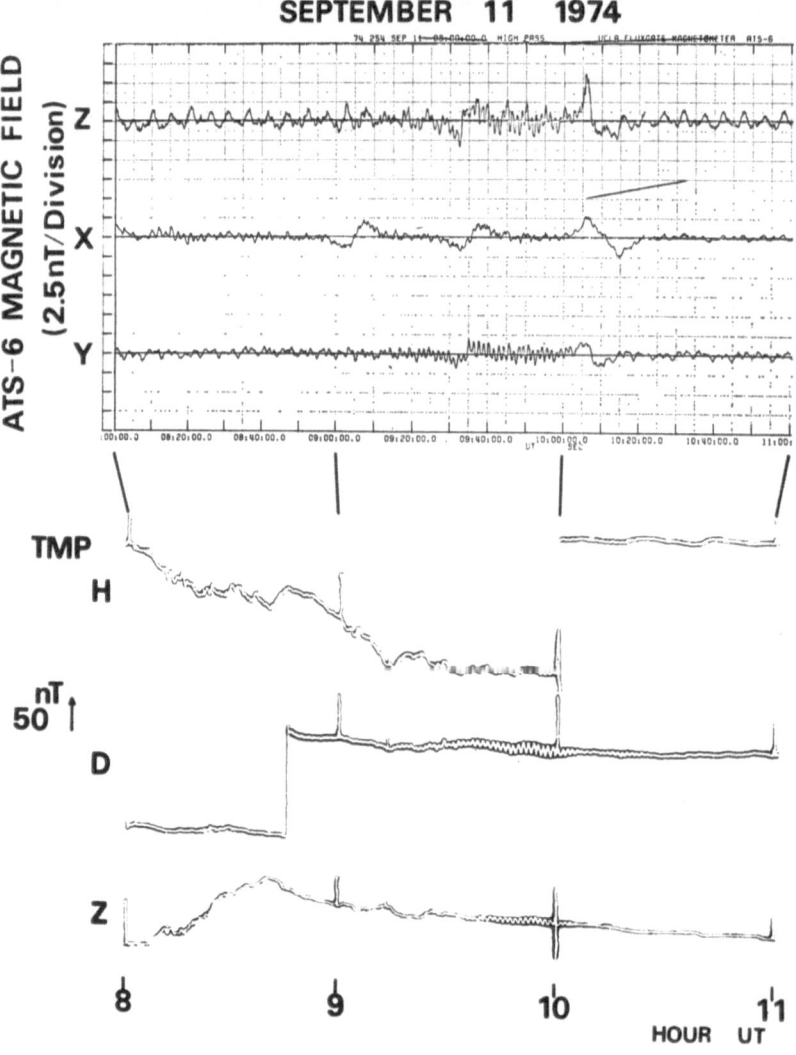

Fig. 13 (a)

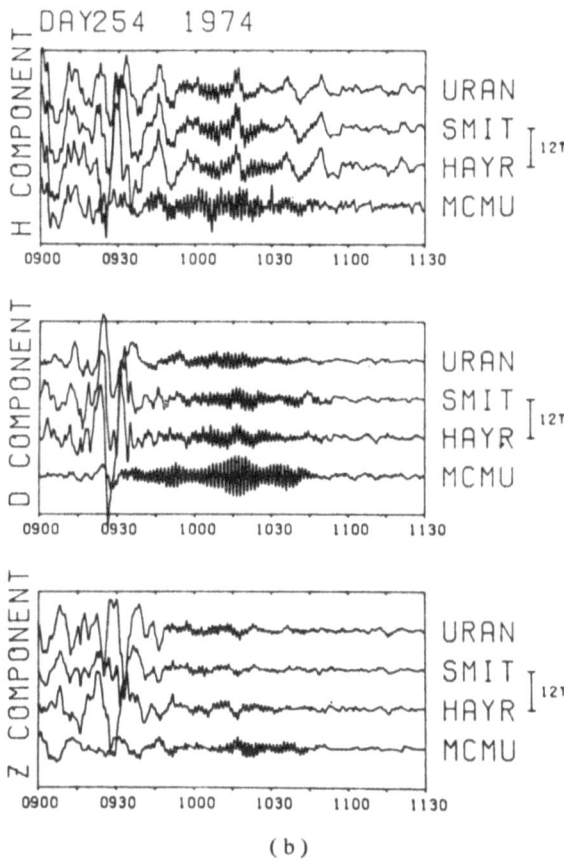

Fig. 13. (a) Simultaneous records of giant pulsations at ATS 6 and Thompson on September
11, 1974.   In this case the Pg wave at ATS 6 appears both in the radial and compressional
components.   (b) Band-pass filtered magnetograms of 1–20 mHz for the September 14,
1974 event observed at Alberta chain stations (after ROSTOKER *et al.*, 1979).

wave on the equatorial plane.   Furthermore, the wave features of Pgs at a latitude of
$\sim 10°$ have also been observed.

Figure 12 shows simultaneous records at ATS 6 and at Thompson on October 5, 1974.
Although the time accuracy of the Thompson record was not enough to correlate each
peak of Pgs in space and on the ground, simultaneous Pg activities with a period of
about 120 sec were clearly seen from 1650 to 1730.   The Pg wave in the magnetosphere
dominated in the radial component than in the azimuthal component and had almost
no compressional component, while the amplitude was largest in the D component on
the ground indicating the usual characteristics of Pgs (e.g. GREEN, 1979).   The amplitude
of wave was as large as about 20 nT on the ground, and was ~ 3 nT in space.   It is interesting to note here that lower period waves seen after 1750 at ATS 6 are virtually not
identified in the ground magnetograms, although the amplitudes of these waves are
approximately the same as that of a Pg.

Another example of simultaneous observations in the magnetosphere and on the

34 S. KOKUBUN

ground is illustrated in Fig. 13. In this case the polarization of the Pg at ATS 6 was different from that of the previous example, and is nearly confined within the magnetic meridian with a significant compressional component, similar to that of the stormtime Pc 5 (BARFIELD *et al.*, 1972). As shown in Fig. 12b, this event was also observed at the University of Alberta magnetometer stations, which were situated in the region approximately 30° west of the ATS 6 meridian (ROSTOKER *et al.*, 1979). They also reported that the event was identified at College, Alaska. This indicates that this event had a longitudinal extent of at least 4 time zones. From the phase difference at longitudinally separated station the azimuthal wave number was estimated as ∼16 for this event (ROSTOKER *et al.*, 1979). It is interesting to note here that pulsations began around 0910 and ended around 1005 at ATS 6 and at Thompson, while they were observed from 0930 to 1050 at the Alberta stations. This suggests that the Pg moves westward during the course of an event.

From the visual scan of magnetograms from Thompson in the 6 month period from July, 1974, 11 Pg events were identified. Seven out of these events were found to occur at ATS 6. It was also noted that the amplitude in space tended to be large in the radial component, indicating a characteristic of the R-class wave.

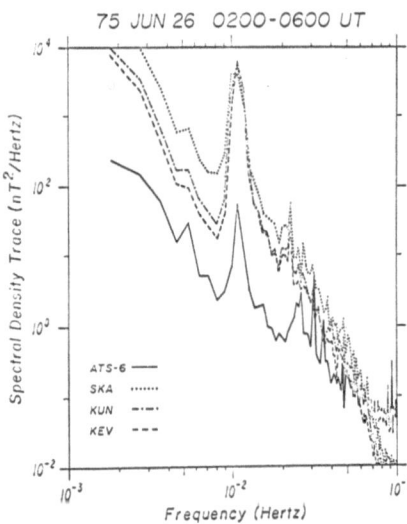

Fig. 14. Power spectra of the Pg event at ATS 6 and at three ground stations, Skarsvaug, Kunes, and Kevo, Norway, on June 26, 1975 (after HILLEBRAND and McPHERRON, 1979).

HILLEBRAND and McPHERRON (1980) have also observed 6 events, which has a strong correlation with ground Pgs. Three of Pgs at equatorial-synchronous altitude were compressional waves, and the others were radially polarized. Figure 14 shows power spectra of the Pg event on June 26, 1975. As clearly in this figure, there are sharp peaks in both the satellite and ground spectra, indicating a strong satellite-ground correlation and a regular wave nature.

5. Simultaneous Satellite-Ground Observations of Pc 3 Waves

Although the characteristics of Pc 3 waves in the magnetosphere have been examined extensively, based on observation at the geostationary orbit (ARTHUR and MCPHERRON, 1975, 1977b; ARTHUR et al., 1977), ground-satellite correlations of these waves are not well understood at present.   DWARKIN et al. (1971) reported in their analysis of data from the almost synchronous satellite, DODGE, at 6.3 R_E in the equatorial plane that transverse and azimuthally oriented Pc 3 oscillations often occurred simultaneously on the ground within a sector of about 60°.   The spectral and polarization analyses of the DODGE data have recently been made by PATEL et al. (1979).   They noted simultaneous Pc 3 activity both in space and on the ground for some cases.

A more extensive study of satellite-ground correlations of Pc 3 waves was made by ARTHUR and MCPHERRON (1977a), using data obtained at ATS 1, at Tungsten (near the foot of the ATS 1 field line), and at College (approximately 600 km west of Tungsten). Although the statistical characteristics of Pc 3 waves are very similar at ATS 1 and at Tungsten (except for the direction of the wave ellipse (ARTHUR et al., 1973; ARTHUR and MCPHERRON, 1975), only one out of 28 Pc 3 events, detected on the ground, could be seen at ATS 1.   The statistical difference of the azimuth of wave polarization agrees with the theoretical inference reported by HUGHES (1974) who concluded that it was the result of propagation through the ionosphere-atmosphere.   For College data, 90 out of 114 events, for which simultaneous data were available from ATS 1, showed simultaneous Pc 3 activity.

As for the simultaneous observation of Pc 3 waves in space and on the ground, a few studies have been made as summarized above.   Further study is need to establish the nature of the ground-satellite correlation of Pc 3 waves.

6.   Discussion

One of important problems in connection with the satellite-ground correlation of ULF waves is to clarify the ionospheric effects on ULF signals from the magnetosphere. NISHIDA (1964) first pointed out the rotation of the major axis of the polarization ellipse in the transmission of the waves through the ionosphere.   Later works, using full-wave numerical solutions to the problem, by INOUE (1973) and HUGHES (1974) indeed showed that the horizontal signal polarization on the ground is at right angles to the transverse magnetospheric signal.   HUGHES and SOUTHWOOD (1976a, b) developed the theory further and reached the following conclusion: the signal polarization ellipse is indeed rotated by the ionosphere: signals with horizontal wave length shorter than the height of the E region, ~ 120 km, are attenuated between the ionosphere and the ground.

As for the rotation of the polarization ellipse, the observations of satellite-ground correlations appear to support this theoretical inference.   As shown previously, the amplitudes of ground-correlated waves in the Pc 4–5 range dominate in the azimuthal component (see Figs. 4, 6, 7, and 8).   KOKUBUN et al. (1976) noted in their polarization analysis of Pc 5 waves in the morning sector, observed at OGO 5, that the orientation of the major axis lies within 20° of the azimuthal direction and is in the first quadrant of the H–D plane (positive H and D) when the polarization hodogram is projected on the northern hemisphere.   Ground-correlated Pc 4 waves, with similar characteristics were also observed near $L \simeq 6.6$ aboard ATS 6, as discussed above.   On the ground the orientation of the major axis of Pc 5 is close to the N–S direction in the second quadrant of the H–D plane at latitude below the demarcation line and only close to the D axis in the second quadrant near the demarcation line (SAMSON, 1972). These facts are consistent with the wave ellipse being rotated by 90° between the magnetosphere and the ground.

A rotation of the wave ellipse is also noted for giant pulsations. GREEN (1979) showed that giant pulsations on the ground were polarized predominantly in the D direction. ATS 6 observations indicate that Pg waves are radially polarized with a significant compressional component. It appears that observational evidence for this type of radial waves is also in favor of the polarization ellipse rotation between the magnetosphere and ground as predicted by theory.

With respect to Pc 3 waves ARTHUR et al. (1977) compared the azimuth-local time variation of the ATS 6–Pc 3 events with that of Pc 3–4 events observed at Lac Rebours ($L \simeq 4$) by LANZEROTTI et al. (1972). This comparison of statistical characteristics also shows the presence of the wave ellipse rotation as expected by theory. However, further reliable simultaneous ground-satellite data are needed to establish the presence or absence of this rotation for Pc 3 waves, since these measurements were made at different L shells.

The differences in degree of satellite-ground correlations appear to reflect the spatial extent of respective disturbances. Azimuthally-polarized waves tend to occur predominately in the morning and around noon. Although compressional giant pulsations are observed in the morning sector, their occurrence is relatively rare as compared with other types of Pc waves. Radially-polarized waves in the Pc 4–5 frequency range most of which have a strong compressional component, generally occur from the early afternoon to midnight (BARFIELD and McPHERRON, 1972b; HEDGECOCK, 1976; HUGHES et al., 1979). It should be important to note that these compressional waves have not yet identified in the ground observation.

As theoretically shown by HUGHES and SOUTHWOOD (1976a, b), the detection of magnetospheric waves on the ground strongly depends on the spatial scale size of disturbances. It appears that the difference in degree of ground-satellite correlations reflect the spatial structures of different modes of waves. OLSON and ROSTOKER (1978) examined the longitudinal phase changes of Pc 4–5 pulsations by using longitudinal station array data. They showed that the azimuthal wave number of ground Pc 4–5 waves is a function of frequency, that is, $m = (1.4 \pm 0.4)f \pm 0.26$ mHz in the frequency range of $1 \sim 12$ mHz. This result indicates that the longitudinal wave length of Pc 4–5 is fairly large. The measured wave numbers clearly showed a change in sign just before noon, but the magnitudes are nearly the same for both morning and afternoon events. Direct measurements of azimuthal characteristics of Pc waves have also been made by HUGHES et al. (1978) by three satellites at synchronous altitude. They have found distinct types of pulsation activity occurring in different local time sectors on the dayside. During the local morning and early afternoon pulsation signals, whose amplitudes predominate in the azimuthal component, are usually coherent over longitude separation of up to 20° and have rather small azimuthal wave number, $m \lesssim 10$. The sign of m changes near local noon, as similar to that of ground observations (OLSON and ROSTOKER, 1978). On the other hand, waves with large radial component occurring after 1500 LT were rarely seen coherently by 2 spacecrafts. This fact indicates a rapid azimuthal variation or large azimuthal wave number.   Also for a compressional Pc 4 event at synchronous altitude near midnight observed by 3 satellites, HUGHES *et al.* (1979) found that the azimuthal wave number was as large as ~ 100 and that the wave was not detected on the ground.

SINGER *et al.* (1979) have also reported simultaneous observations of Pc 3, 4, and 5 pulsations by 5 satellites in the pre-noon time sector at and near synchronous orbit. The magnetic field oscillations, examined by SINGER *et al.* (1979), were dominant in the component transverse to the ambient magnetic field and directed radially from the earth. They found that the periods of these simultaneous pulsations are not the same at the different observation points, indicating the spatially limited nature of radial waves. Thus, these in situ observations indicate that the spatial extent of R-class Pc waves is more limited than that of A-class waves.   This tendency seems to be supported by fact that the azimuthal wave number of ground Pc 4–5, observed by OLSON and ROSTOKER (1978), is not so large, since the ground-correlated Pc 4–5 waves are mostly polarized in the azimuthal direction.   Therefore, a lack of satellite-ground correlation for R-class Pc 4–5 waves appears to reflect the spatially limited nature of these waves.

The author would like to thank R. L. McPherron for providing ATS 6 data and for discussions.

REFERENCES

ARTHUR, C. W., Pc 4 magnetic pulsations at synchronous orbit, ATS-6, *EOS*, **59**, 1166, 1978.

ARTHUR, C. W. and R. L. MCPHERRON, Micropulsations in the morning sector, 2.   Satellite observations of 10- to 45-s waves at synchronous orbit, ATS 1, *J. Geophys. Res.*, **80**, 4621–4626, 1975.

ARTHUR, C. W. and R. L. MCPHERRON, Micropulsations in the morning sector, 3.   Simultaneous ground-satellite observations of 10- to 45-s period waves near $L=6.6$, *J. Geophys. Res.*, **82**, 2859–2866, 1977a.

ARTHUR, C. W. and R. L. MCPHERRON, Interplanetary magnetic field conditions associated with synchronous orbit observations of Pc 3 magnetic pulsations, *J. Geophys. Res.*, **82**, 5138–5142, 1977b.

ARTHUR, C. W., R. L. MCPHERRON, and P. J. COLEMAN, Jr., Micropulsations in the morning sector, 1. Ground observations of 10- to 45-s waves, Tungsten, Northwest Territories, Canada, *J. Geophys. Res.*, **78**, 8180–8192, 1973.

ARTHUR, C. W., R. L. MCPHERRON, and W. J. HUGHES, A statistical study of Pc 3 magnetic pulsations at synchronous orbit, ATS 6, *J. Geophys. Res.*, **82**, 1149–1157, 1977.

BARFIELD, J. N. and R. L. MCPHERRON, Investigation of interaction between Pc 1 and 2 and Pc 5 micropulsations at the synchronous equatorial orbit, *J. Geophys. Res.*, **77**, 4704–4719, 1972a.

BARFIELD, J. N. and R. L. MCPHERRON, Statistical characteristics of storm associated Pc 5 micropulsations observed at the synchronous equatorial orbit, *J. Geophys. Res.*, **77**, 4720–4733, 1972b.

BARFIELD, J. N. and R. L. MCPHERRON, Stormtime Pc 5 magnetic pulsations observed at synchronous orbit and their correlation with the partial ring current, *J. Geophys. Res.*, **83**, 739–743, 1978.

BARFIELD, J. N., R. L. MCPHERRON, P. J. COLEMAN, Jr., and D. J. SOUTHWOOD, Storm associated Pc 5 micropulsation events observed at the synchronous equatorial orbit, *J. Geophys. Res.*, **77**, 143–158, 1972.

BARFIELD, J. N., L. J. LANZEROTTI, C. G. MACLENNAN, G. A. PAULIKAS, and M. SCHULZ, Quiet time observation of a coherent compressional Pc-4 micropulsation at synchronous altitude, *J. Geophys. Res.*, **76**, 5252–5258, 1971.

BOSSEN, M., R. L. MCPHERRON, and C. T. RUSSELL, Simultaneous Pc 1 observations by the synchronous satellite ATS-1 and ground stations: Implications concerning IPDP generation mechanisms, *J. Atmos. Terr. Phys.*, **38**, 1157–1167, 1976a.

BOSSEN, M., R. L. MCPHERRON, and C. T. RUSSELL, A statistical study of Pc 1 magnetic pulsations at synchronous orbit, *J. Geophys. Res.*, **81**, 6083–6091, 1976b.

CHEN, L. and A. HASEGAWA, A theory of long-period magnetic pulsations, 1. Steady state excitation of field line resonances, *J. Geophys. Res.*, **79**, 1024–1032, 1974.

CUMMINGS, W. D., S. E. DEFOREST, and R. L. MCPHERRON, Measurements of the Poynting vector of standing hydromagnetic waves at geosynchronous orbit, *J. Geophys Res.*, **83**, 697–706, 1978.

CUMMINGS, W. D., R. J. O'SULLIVAN, and P. J. COLEMAN, Jr., Standing Alfvén waves in the magnetosphere, *J. Geophys. Res.*, **74**, 778–793, 1969.

CUMMINGS, W. D., C. COUTEE, D. LYONS, and W. WILEY III, The dominant mode of standing Alfvén waves at the synchronous orbit, *J. Geophys. Res.*, **80**, 3705–3708, 1975.

DEFOREST, S. E. and C. E. MCILWAIN, Plasma clouds in the magnetosphere, *J. Geophys. Res.*, **76**, 3578–3611, 1971.

DWARKIN, M. L., A. J. ZMUDA, and W. E. RADFORD, Hydromagnetic waves at 6.25 earth's radii with periods between 3 and 240 seconds, *J. Geophys. Res.*, **76**, 3668–3674, 1971.

GREEN, C. A., Observations of Pg pulsations in the northern auroral zone and at lower latitude conjugate regions, *Planet. Space Sci.*, **27**, 63–77, 1979.

HEDGECOCK, P. C., Giant Pc 5 pulsations in the outer magnetosphere: A study of HEOS-1 data, *Planet. Space Sci.*, **24**, 921–935, 1976.

HIGBIE, P. R., R. D. BELIAN, and D. N. BAKER, High-resolution energetic particle measurements at 6.6 R_E, 1. Electron micropulsations, *J. Geophys. Res.*, **83**, 4851–4855, 1978.

HILLEBRAND, O. and R. L. MCPHERRON, Simultaneous observations of magnetic Pc 4 pulsations at the synchronous orbit and on the ground, to be published in *J. Geophys. Res.*, 1980.

HUGHES, W. J., The effect of the atmosphere and ionosphere on long period magnetospheric micropulsations, *Planet. Space Sci.*, **22**, 1157–1172, 1974.

HUGHES, W. J. and D. J. SOUTHWOOD, The screening of micropulsation signals by the atmosphere and ionosphere, *J. Geophys. Res.*, **81**, 3234–3240, 1976a.

HUGHES, W. J. and D. J. SOUTHWOOD, An illustration of modification of geomagnetic pulsation structure by the ionosphere, *J. Geophys. Res.*, **81**, 3241–3247, 1976b.

HUGHES, W. J., R. L. MCPHERRON, and J. N. BARFIELD, Geomagnetic pulsations observed simultaneously on three geostationary satellites, *J. Geophys. Res.*, **83**, 1109–1116, 1978.

HUGHES, W. J., R. L. MCPHERRON, J. N. BARFIELD, and B. H. MAUK, A compressional Pc 4 pulsation observed by three satellites in geostationary orbit near local midnight, *Planet. Space Sci.*, **26**, 821–839, 1979.

INOUE, Y., Wave polarizations of geomagnetic pulsations observed in high latitudes on the earth's surface, *J. Geophys. Res.*, **78**, 2959–2976, 1973.

KAYE, S. M. and M. G. KIVELSON, Observations of Pc 1–2 waves in the outer magnetosphere, *J. Geophys. Res.*, **84**, 4267–4276, 1979.

KOKUBUN, S. and T. NAGATA, Geomagnetic pulsations Pc 5 in and near the auroral zones, *Rep. Ionos. Space Res. Japan*, **22**, 158–176, 1965.

KOKUBUN, S., R. L. MCPHERRON, and C. T. RUSSELL, Ogo 5 observations of Pc 5 waves: Ground-magnetosphere correlations, *J. Geophys. Res.*, **81**, 5141–5149, 1976.

KOKUBUN, S., M. G. KIVELSON, R. L. MCPHERRON, C. T. RUSSELL, and H. I. WEST, Jr., Ogo 5 observations of Pc 5 waves: Particle flux modulations, *J. Geophys. Res.*, **82**, 2774–2786, 1977.

LANZEROTTI, L. J. and H. FUKUNISHI, Modes of magnetohydrodynamic waves in the magnetosphere, *Rev. Geophys. Space Phys.*, **12**, 724–729, 1974.

LANZEROTTI, L. J. and D. J. SOUTHWOOD, Hydromagnetic waves, in *Solar System Plasma Physics*, Vol. 3, edited by C. F. Kennel, L. J. Lanzerotti, and E. N. Parker, pp. 111–135, North-Holland Pub., 1979.

LANZEROTTI, L. J. and N. A. TARTAGLIA, Propagation of a magnetospheric compressional wave to the ground, *J. Geophys. Res.*, **77**, 1934–1940, 1972.

LANZEROTTI, L. J., A. HASEGAWA, and N. A. TARTAGLIA, Morphology and interpretation of magnetospheric plasma waves at conjugates points during December solstice, *J. Geophys. Res.*, **77**, 6731–6745, 1972.

LANZEROTTI, L. J., H. FUKUNISHI, C. C. LIN, and L. J. CAHILL, Jr., Storm time Pc 5 magnetic pulsation at the equator in the magnetosphere and its latitude dependence as measured on the ground, *J. Geophys.*

Res., **79**, 2420–2425, 1974.

LANZEROTTI, L. J., C. G. MACLENNAN, H. FUKUNISHI, J. K. WALKER, and D. J. WILLIAMS, Latitude and longitude dependence of storm time Pc 5 type plasma wave, *J. Geophys. Res.*, **80**, 1014–1018, 1975.

LIN, C. C. and L. J. CAHILL, Jr., Pc 4 and Pc 5 pulsations during storm recovery, *J. Geophys. Res.*, **81**, 1751–1761, 1976.

McPHERRON, R. L., Description of the UCLA fluxgate magnetometer on ATS-6: Instrument, data files, data displays, preliminary observations, IGPP publication No. 1578, UCLA, May 19, 1976.

McPHERRON, R. L., P. J. COLEMAN, Jr., and R. C. SNARE, ATS-6/UCLA fluxgate magnetometer, *IEEE Trans. Aerosp. Electron. Syst.*, AES-11, 1110–1117, 1975.

McPHERRON, R. L., C. T. RUSSEL, and P. J. COLEMAN, Jr., Fluctuating magnetic fields in the magnetosphere, 2. ULF waves, *Space Sci. Rev.*, **13**, 411–454, 1972.

NISHIDA, A., Ionospheric screening effect and storm sudden commencement, *J. Geophys. Res.*, **69**, 1861–1874, 1964.

OLSON, J. V. and G. ROSTOKER, Longitudinal phase variations of Pc 4–5 micropulsations, *J. Geophys. Res.*, **83**, 2481–2488, 1978.

PATEL, V. L., Low frequency hydromagnetic waves in the magnetosphere: Explorer XII, *Planet. Space Sci.*, **13**, 485–506, 1965.

PATEL, V. L. and L. J. CAHILL, Jr., Evidence of hydromagnetic waves in the earth's magnetosphere and of their propagation to the earth's surface, *Phys. Rev. Lett.*, **12**, 213–215, 1964.

PATEL, V. L., R. J. GREAVES, S. A. WAHAB, and T. A. POTEMRA, Dodge satellite observations of Pc 3 and Pc 4 magnetic pulsations and correlated effects in the ground observations, *J. Geophys. Resl*, **84**, 4257–4266, 1979.

ROSTOKER, G., H.-L. LAM, and J. V. OLSON, Pc 4 giant pulsations in the morning sector, *J. Geophys. Res.*, **84**, 5153–5166, 1979.

SAMSON, J. C., Three-dimensional polarization characteristics of high latitude Pc 5 geomagnetic pulsations, *J. Geophys. Res.*, **77**, 6145–6160, 1972.

SAMSON, J. C., J. A. JACOBS, and G. R. ROSTOKER, Latitude dependent characteristics of long-period geomagnetic pulsations, *J. Geophys. Res.*, **76**, 3675–3683, 1971.

SINGER, H. J. and M. G. KIVELSON, The latitudinal structure of Pc 5 waves in space: Magnetic and electric field observations, *J. Geophys. Res.*, **84**, 7213–7222, 1979.

SINGER, H. J., C. T. RUSSELL, M. G. KIVELSON, T. A. FRITZ, and W. LENNARTSSON, The spatial extent and structure of Pc 3, 4, 5 pulsations near the magnetospheric equator, *Geophys. Res. Lett.*, **6**, 889–892, 1979.

SOUTHWOOD, D. J., Some features of field line resonances in the magnetosphere, *Planet. Space Sci.*, **22**, 483–491, 1974.

SOUTHWOOD, D. J. and W. J. HUGHES, Source induced vertical components in geomagnetic pulsation signals, *Planet. Space Sci.*, **26**, 715–720, 1978.

SU, S.-Y., A. KONRADI, and T. A. FRITZ, On propagation direction of ring current proton ULF waves observed ATS 6 at 6.6 R_E, *J. Geophys. Res.*, **82**, 1859–1868, 1977.

SU, S.-Y., A. KONRADI, and T. A. FRITZ, On energy dependent modulation of the ULF ion flux oscillations observed at small pitch angles, *J. Geophys. Res.*, **84**, 6510–6516, 1979.

Multisatellite Observations of Geomagnetic Pulsations

W. J. HUGHES

Department of Astronomy, Boston University, Boston, Massachusetts, U.S.A.

(Received June 28, 1980)

The technique of using simultaneous data from more than one spacecraft to study hydromagnetic waves in space is proving to be a very powerful one which will undoubtedly become more widespread in the future. We review the results obtained using this technique thus far. All of the work has involved longer period (>20 s) pulsations and most of the data has come from near geostationary orbit. Attempts to measure the thickness of field line resonance regions have not so far been conclusive, but they suggest that values of about $(1/2)R_E$ are typical. Signals in the Pc 3 and 4 bands in the morning are coherent over at least 20° of longitude while in the afternoon they are much more localized. One afternoon source is resonance with hot ions. Pi 2's on the nightside can also be confined to small regions of longitude.

1. Introduction

Our understanding of the longer period geomagnetic pulsations increased dramatically about ten years ago due to two advances in observational technique. For the first time magnetometers sensitive enough to record ULF waves in space were flown on spacecraft (e.g. ATS 1; CUMMINGS *et al.*, 1969). These observations showed incontrovertibly the magnetospheric nature of pulsations and soon led to many types of waves being identified in space. Second, large arrays of closely spaced ground based magnetometers designed for pulsation research began to be deployed (e.g. SAMSON *et al.*, 1971; FUKUNISHI and LANZEROTTI, 1974). The data from these arrays allowed us to get a much better picture of the spatial variations in pulsation signals and it was these results which stimulated the theoretical studies which led to detailed descriptions of field line resonance structure (SOUTHWOOD, 1974; CHEN and HASEGAWA, 1974). A result of this has been the close interaction between theory and observation which still continues, the renewed interest in pulsation research and the big advances we have made over the last few years which are all witnessed by this symposium.

More recently the increasing number of scientifically equipped satellites have made it possible to combine these two techniques. By using several spacecraft with similar magnetometers aboard we can create a de facto magnetometer array in space. This gives us the advantages of a station network without the disadvantage of an intervening ionosphere between the observer and the wave source. The original studies using this technique relied on chance close encounters between spacecraft, particularly geosynchronous or near geosynchronous ones. The most recent work uses the ISEE spacecraft, a trio of spacecraft designed to provide intersatellite correlations, two of which follow each other closely in the same orbit so are ideally suited for this work. The first ULF wave studies using ISEE data are just coming into print and will be described later in this paper.

This paper attempts to review the work which has used more than one spacecraft to obtain simultaneous observations of ULF pulsations and to summarize the new results that have come out of such studies.   In this sense this is a technique oriented rather than a phenomenon oriented paper.   It is however a technique which has shown itself to be a very powerful tool for pulsation work, especially when electric field or particle data from the spacecraft is used in conjunction with the more traditional magnetic field data. An added advantage of gathering data in the magnetosphere is that this extra information is often available.   Multiple satellite observation is a technique that is rapidly ceasing to be novel and is already almost a standard way to study ULF pulsations.   Moreover, it has the potential to stimulate the field over the next decade in much the same way as ground array and single spacecraft studies have done for the last ten years or so.

All the work described here concerns longer period pulsations of the various types (Pc 3, 4, and 5 and Pi 2) and comes mostly from near geostationary orbit.   The paper is organized loosely by local time.   We will describe the waves seen in the outer magnetosphere at different local times starting near dawn, going on past noon towards dusk and finally ending near midnight, concentrating on describing those new results which have come from multiple satellite studies.   But before starting this tour I will first discuss observations of pulsation resonant regions, a topic common to most of the longer period types of pulsation.   This topic also allows us to see how multiple satellite studies are maturing.

2.   The Resonant Region

DUNGEY (1954, see also DUNGEY, 1963) first suggested that pulsations may be hydromagnetic waves resonating on geomagnetic field lines.   However it wasn't until the late sixties that latitudinal chains of magnetometers (e.g. SAMSON et al., 1971) showed how important this idea was.   The data from the chains showed that the amplitude of a particular pulsation signal peaks at some latitude (or L value).   The latitude at which this peak occurs depends on the pulsation period; shorter period waves peak at lower latitudes. The width of this peak in latitude or L value gives a measure of the amount of damping the signal is experiencing, or the quality factor (Q) of the resonance.   Unfortunately one of the effects of the ionosphere is to smooth spatial variations of signals seen on the ground (HUGHES and SOUTHWOOD, 1976).   No detail can be seen on the ground which has scale lengths less than about 120 km, the height of the E region, as the magnetic fields seen on the ground are the result of Hall currents flowing in the E region.   So there is a fundamental limit to the narrowness of a peak as measured on the ground and it is infuriatingly near typical peak widths.   As a result, definitive measurements can only be made in space or in the ionosphere (e.g. WALKER et al., 1979).

Results in space are few because the spacecraft must be close together to measure the short scale lengths we expect.   The first attempt was made by HUGHES et al. (1977) who used data from OGO 5 and ATS 1.   These spacecraft both saw a Pc 4 pulsation near local noon.   By using differences in the polarization of the signal at both spacecraft, the authors concluded that the resonance region lay on an L shell between the two satellites. But as the spacecraft never approached each other closely, all that could be obtained

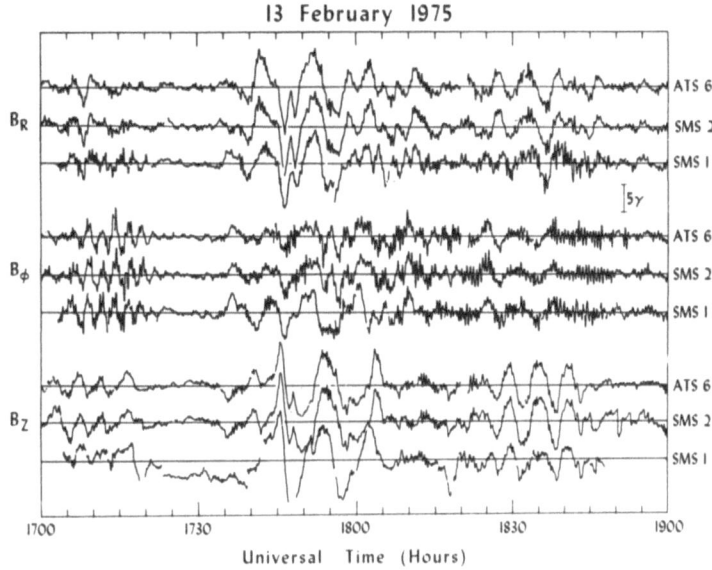

Fig. 1. Wave forms in the magnetic field measured by three geostationary spacecraft. The spacecraft positions are: ATS 6 and SMS 2, 95°W, SMS 1, 75°W. The data has been high pass filtered to remove long period (>10 min) trends (HUGHES *et al.*, 1978a).

was an upper estimate of the resonant region thickness which was about 1 R_E.

A better estimate was made using data from ATS 6 and SMS 2 (HUGHES *et al.*, 1978a) which passed within 600 km of each other on 13 February 1975. Data from the interval when they passed is shown in Fig. 1 together with data from SMS 1 which was 20° to the East. The satellites were again close to local noon. Three different wave periods are visible in the figure, a long period (500 s) Pc 5, largely confined to the meridianal (R–Z) plane, and both Pc 4 ($\tau \sim 120$ s) and Pc 3 ($\tau \sim 30$ s) waves best visible in the azimuthal component (B_ϕ). Figure 2 shows spectra made from the central hour of this interval. The top panel shows autospectra of the radial (B_R) and azimuthal components from both ATS 6 and SMS 2. Below are shown the coherence between these signals and the phase difference between them as a function of frequency. The radial components are coherent and in phase at all frequencies. The azimuthal components are coherent only over two narrow frequency bands corresponding to the Pc 4 and Pc 3 wave signals as is seen by the peaks in the top panel. Where they are coherent, the azimuthal components are also not in phase, but rather, ATS 6, the satellite closest to the Earth, leads in phase. Now theory predicts (e.g. SOUTHWOOD, 1975) and ground based and ionospheric observations support the fact that in the magnetosphere the radial magnetic component is in phase across a resonance region, but that the azimuthal component undergoes a 180° phase shift with the smaller L values leading in phase. This is just what we see here. The spacecraft were about 0.1 R_E apart radially. Resonance region widths of about 0.5 and 0.25 R_E for the Pc 4 and Pc 3 signals respectively were deduced from the phase differences shown. At these L values, field lines 0.5 R_E apart at the equator are about 100 km apart on the ground, just about the spatial resolution limit for ground based measurements,

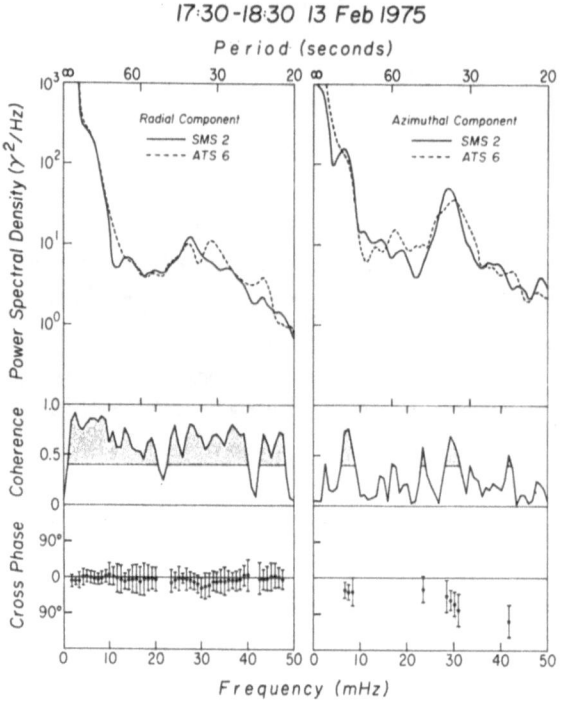

Fig. 2. Spectra of the magnetic data from SMS 2 and ATS 6 from the interval when SMS 2
drifted closest to ATS 6. SMS 2 and ATS 6 were separated radially, about 600 km
apart (HUGHES *et al.*, 1978a).

These thicknesses are consistant with both the observed damping time scales of pulsations, and estimated ionospheric damping rates (NEWTON *et al.*, 1978; ALLAN and KNOX, 1979).

Such close satellite spacing occurs only very rarely by chance.   However, the ISEE 1 and ISEE 2 spacecraft provide this spacing routinely and are never much more than $1/2\ R_E$ apart in the middle magnetosphere.   Figure 3 is taken from a recent study by SINGER *et al.* (1979) and shows simultaneous pulsation events from five spacecraft whose positions are shown inset.   This is a remarkable collection of simultaneous spacecraft data of ULF waves all from within a four hour local time sector.   The data shows that a wide range of pulsations can exist simultaneously in closely neighboring regions of the magnetosphere. The three geostationary spacecraft show continuous wave trains varying in period from 75 s to 120 s.   The data from the ISEE spacecraft is worth closer study and is redrawn in Fig. 4.   The ISEE satellites were inbound with ISEE 2 about 5 min ($\sim 1/4\ R_E$) ahead of ISEE 1 in the orbit.   In the top panel are plotted simultaneous observations from both spacecraft versus U.T.   As the spacecraft approach the Earth the wave frequency seen by both gets progressively higher which suggests that we are always seeing waves at or close to the local resonant frequency.   This is borne out by the lower panel in which the data has been replotted with observations at a given spatial location plotted versus L value.   In this panel there is a much better correspondence between the periods seen by the two spacecraft than in the upper panel indicating that the spacecraft are passing through

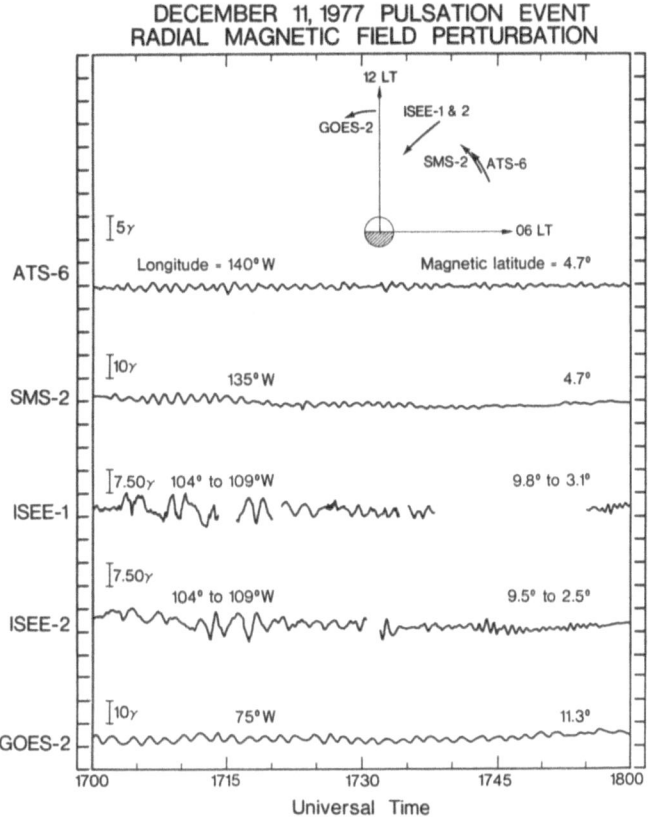

Fig. 3. Wave forms in the radial component of the magnetic field measured by five spacecraft whose positions are shown in the inset (SINGER *et al.*, 1979).

a spatial structure.   There is an annoying data gap in the ISEE 1 record, but the sudden
cessation of activity at $L \simeq 4.5$ is clear at both spacecraft.   Because of the simultaneous
excitation of several resonances it is difficult to get a definitive measurement of resonance
thickness from this pass.   However, Fig. 4 shows that the scale length is certainly less than
1 R_E.   Hopefully other passes, particularly when the spacecraft are closer together, will
provide more definitive information.   But the occurrence of these several simultaneous
resonances means that either the wave source is very broad band or that there are many
localized wave sources.

3.   Morning Pc 3 and 4 Waves

One of the first goals of multispacecraft studies was to determine east/west wave
lengths of pulsation signals using geostationary spacecraft.   This parameter helps distinguish between various wave generation modes and the first ground based measurements
showed a somewhat confused picture, especially at mid latitudes (HERRON, 1966; GREEN,
1976; OLSON and ROSTOKER, 1978; MIER-JEDRZEJOWICZ and SOUTHWOOD, 1979).

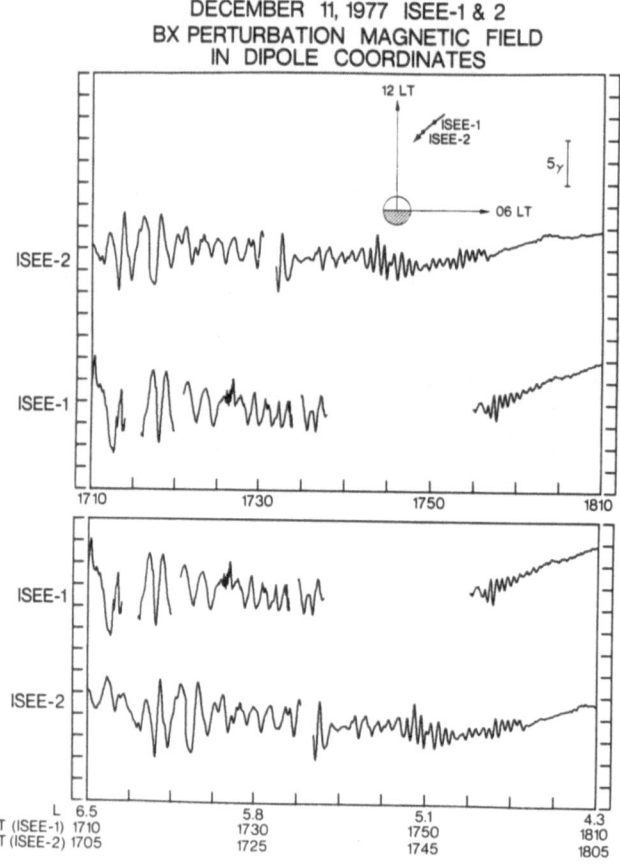

Fig. 4. Magnetic data from ISEE 1 and 2 for the same interval as Fig. 3. In the top panel observations made at the same time are plotted above each other. In the bottom panel observations made at the same place are plotted above each other (Singer *et al.*, 1979).

The results of computing the azimuthal wave or *m* number (phase difference per degree of longitude) from a week's data from ATS 6 and SMS 2 around the time they were closest together is shown in Fig. 5. The results are plotted versus U.T. with local noon shown as a dashed line. The data was divided into three period bands. In each band during the local morning hours the wave number is predominantly negative indicating westward propagation. Around noon the sign tends to change. This is especially clear in the longest period band corresponding to Pc 4, and now indicates eastward propagation, again away from the noon meridian. Nearly all the waves seen at the two spacecraft in the morning hours during this week were coherent enough to yield a data point in Fig. 5. The lack of points in the afternoon hours is not due to a lack of wave events but rather to a lack of any waves coherent between the spacecraft indicating that afternoon waves are of a very different type. This result was unexpected.

There were two dominant wave periods seen during this week in the morning hours. One set had periods in the Pc 3 band close to 30 s. Waves of this type have been studied

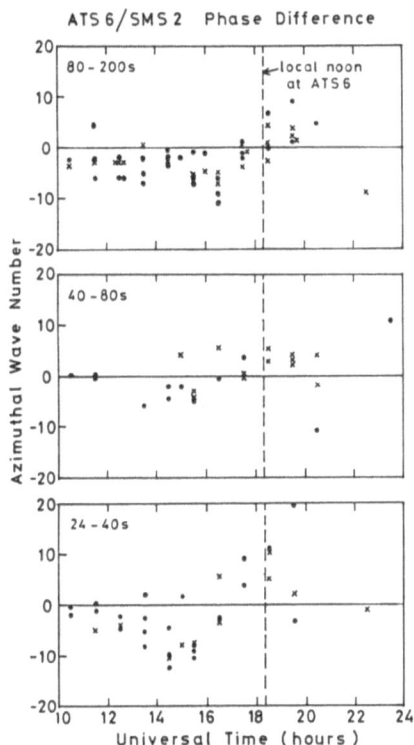

Fig. 5. The azimuthal wave number, m, obtained for waves which were coherent between SMS 1 and ATS 6, plotted as a function of universal time. Dots and crosses refer to values obtained from azimuthal and radial components respectively.

in detail by ARTHUR *et al.* (1977) and it has recently been shown that their occurrence is closely controlled by solar wind parameters (Takahashi and McPherron, private communication). What multiple satellite studies have shown is that they are often coherent over fairly long E–W separations, up to 20°, and that they appear to propagate predominantly westward, with a tendency for eastward propagation around noon.

The other dominant period was in the upper range of the Pc 4 band 120–150 s. Figure 6 shows spectra of data from three geostationary spacecraft at a time when the intersatellite spacings were 12° and 7° between SMS 1/SMS 2 and SMS 2/ATS 6 respectively. These are typical mid morning spectra with peaks at about 150 s and 40 s and are not dissimilar to Fig. 2. The peak at 150 s shows best in the azimuthal component spectra and the signals are coherent between all three spacecraft. The phase differences yield a consistent m number of about -2. Morning Pc 4 are typically azimuthally polarized, and coherent between the spacecraft with small negative m numbers. The propagation direction switches near noon.

At first glance this appears strong evidence for the now classical picture of Kelvin-Helmholtz driven waves resonating on field lines deep in the magnetosphere. But there are a couple of annoying discrepancies. One is the extremely low wave numbers. An m number of -2 for a 150 s wave yields a phase velocity at the magnetopause of about 1,800 km/s, far in excess of even the maximum solar wind speeds. A second point is brought out in Fig. 7 which shows part of an azimuthally polarized Pc 4 event ($\tau \simeq 120$ s)

14:25–15:25 12 Feb 1975

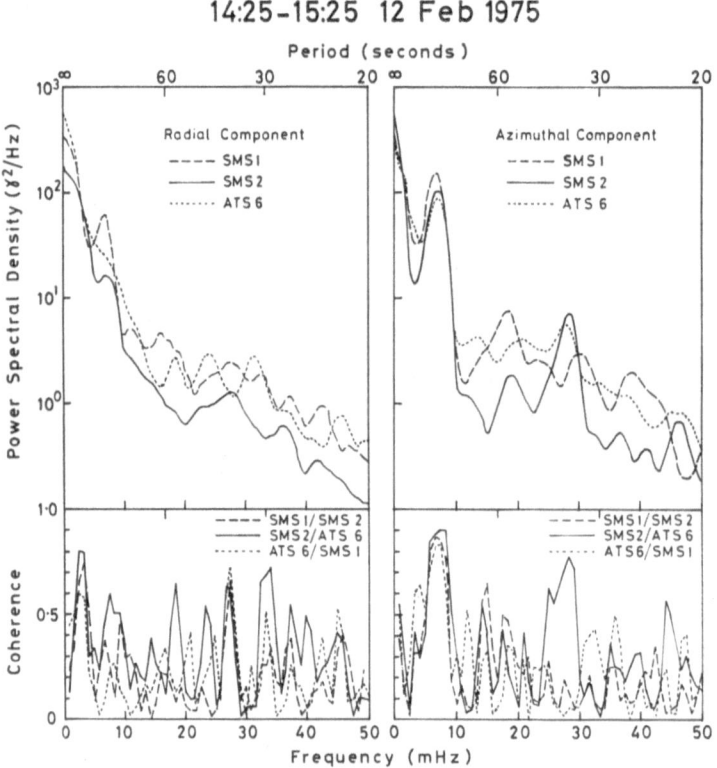

Fig. 6.   Spectra of the magnetic variations measured by three geostationary spacecraft at
a local time of about 1000 (HUGHES *et al.*, 1978a).

which occurred near 0800 LT.   Data from two spacecraft is shown superimposed.   It
is immediately apparent that the waves at the two spacecraft have slightly different frequencies.   Inspecting the B_ϕ component in particular we see that at 12:45 as a new wave
packet starts the oscillations at both spacecraft are in phase.   But the peaks at SMS 2
come progressively sooner than at ATS 6 so that by 12:55 they are approximately half
a period out of phase.   By then the wave amplitude has decayed also.   Then a new packet
starts, again in phase, and again the signals drift slowly out of phase.   This behavior
is emphasized in the lowest part of the figure. Δt is the time between a peak in the B_ϕ
trace at SMS 2 and the next peak at ATS 6.   The dots fall along a characteristic ramp
shape, the sloped portions occurring during wave packets as the signals drift apart in
phase and the step occurring at the start of new wave packets when sudden phase shifts
bring the signals back into phase.   This would be explained if an impulse, traveling at
the Alfvén speed, say, were to set off oscillations almost simultaneously everywhere.
Thus wave packets would start in phase everywhere, but would then oscillate at the local
resonant frequency which differs slightly from place to place, causing the signals to drift
apart in phase.   The next impulse would start a new packet again in phase at all stations.
    Both these points go against the Kelvin–Helmholtz picture.   They might not be
unrelated however.   The *m* numbers were calculated from hour long segments of data,

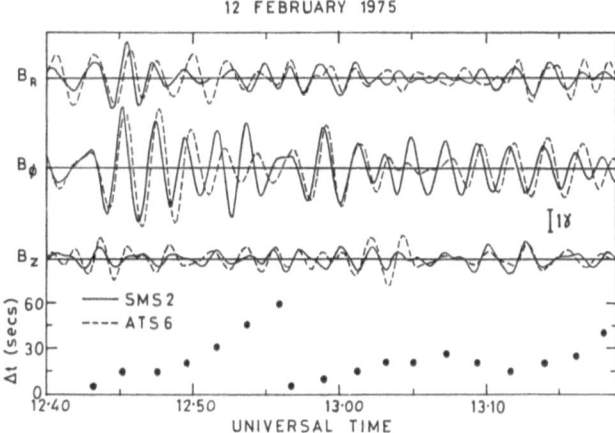

Fig. 7. Magnetic field data from two geostationary spacecraft. SMS 2 (solid lines) and
ATS 6 (dashed lines). The data is presented in the spacecraft coordinate system and
has been filtered with a passband between 5 and 12 mHz. The dots at the bottom of
the figure show the time interval between a maximum B_ϕ at SMS 2 and the next maximum
at ATS 6. Local time is about 0700.

which would typically contain many wave packets. If the signals are not coherent from
packet to packet as Fig. 7 suggests, the m number derived would be some sort of averaged
m number. Clearly more work needs to be done on the packet structure of Pc's.

4.   Afternoon Pc 4's

The week in February 1975 when ATS 6 and SMS 1 and 2 were close was a rather
disturbed period. We have already mentioned that the lack of points in the afternoon
hours in Fig. 5 was due to lack of coherent waves. Figure 8 shows Pc 4 period waves
seen by three spacecraft near 1500 LT. In contrast to the morning events these waves are
radially polarized and they have a less clearly sinusoidal wave form. Note, too, that
the waves stop and start independently at the different spacecraft. Figure 9 shows spectra
of the second hour's data in Fig. 8 and contrasts strongly with Fig. 6. The larger peaks
are now in the radial component and at different frequencies at the different spacecraft,
while the coherence values are small. The different spacecraft are seeing different waves.
Even though the region where the waves are occurring is large the waves themselves are
localized. In analyzing plasma data from ATS 6, HUGHES et al. (1978b) found that at
this time the ion distribution had a bump in tail form, as is shown in Fig. 10. The slope
of the energy distribution is positive between energies of 1 and 10 keV. A 3 keV proton
at these L values has a bounce period of about 100 s. The wave in the magnetic field data
had a period of 90 s. Detailed analysis of the plasma data showed that the wave electric
field which caused the low energy protons to drift was in quadrature with the magnetic
field, indicating a standing wave structure. The phase difference also indicated that it
was an even harmonic (probably the second) with an electric node at the equator. This
is crucial, for particles can only bounce resonate with even harmonic waves (SOUTHWOOD

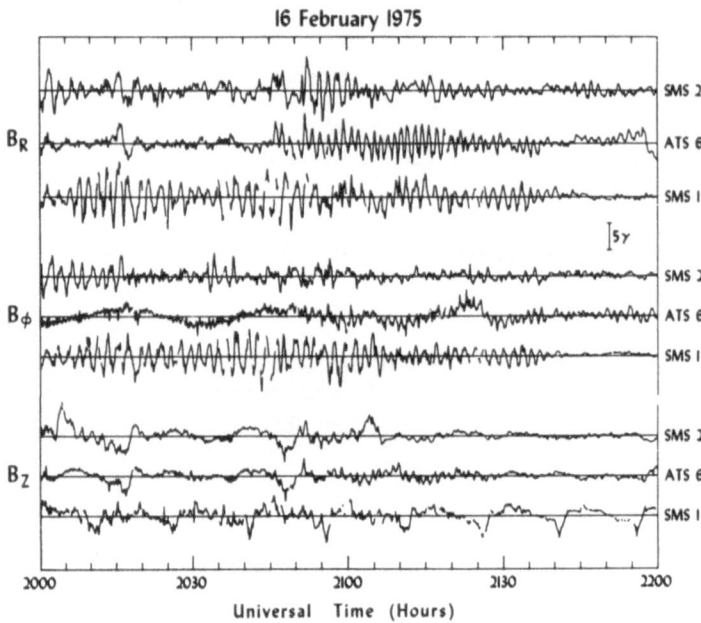

Fig. 8. Pc 4 waves seen in the magnetic field data from SMS 1, ATS 6, and SMS 2, at about 1500 LT. Spacecraft positions are: SMS 1, 75°W, ATS 6, 95°W, and SMS 2, 105°W. The data has been high pass filtered with a cut off at a period of 10 min.

et al., 1969) whereas most other wave sources produce odd harmonics.   Taken altogether this seems to be clear evidence for waves produced by bounce resonance with hot protons. Distributions of the type shown in Fig. 10 are not atypical of the late afternoon at geostationary orbit.   COWLEY (1976) has shown how convection of particles from the tail leads to such distributions.   Further circumstantial evidence is that Troitskaya (this symposium and private communication) has found that the occurrence of similar Pc 4 period waves at ATS 6 is associated with observations of IPDP type pulsations on the ground.   IPDP are believed to be generated by gyroresonance with hot ions.

The generation of these waves can be pictured as analogous to a laser mechanism. The field line acts as the resonant cavity with a mirroring ionosphere at each end allowing only some of the radiation to escape.   The inverted particle distribution is provided by an energy pump in the substorm process and reverts to an ordinary distribution by giving off energy coherently in exciting the hydromagnetic wave.

5. Nighttime Pulsations

There has not yet been much systematic study of nighttime pulsations using multi-satellite data sets.   An isolated event was studied by HUGHES *et al.* (1979) and provided the first measurement of a large m number.   The magnetic data showed that this event was highly monochromatic with a period of about 55 s.   It was seen throughout a three hour local time sector just prior to local midnight.   However the east/west wavelength

21:00-22:00 16 Feb 1975

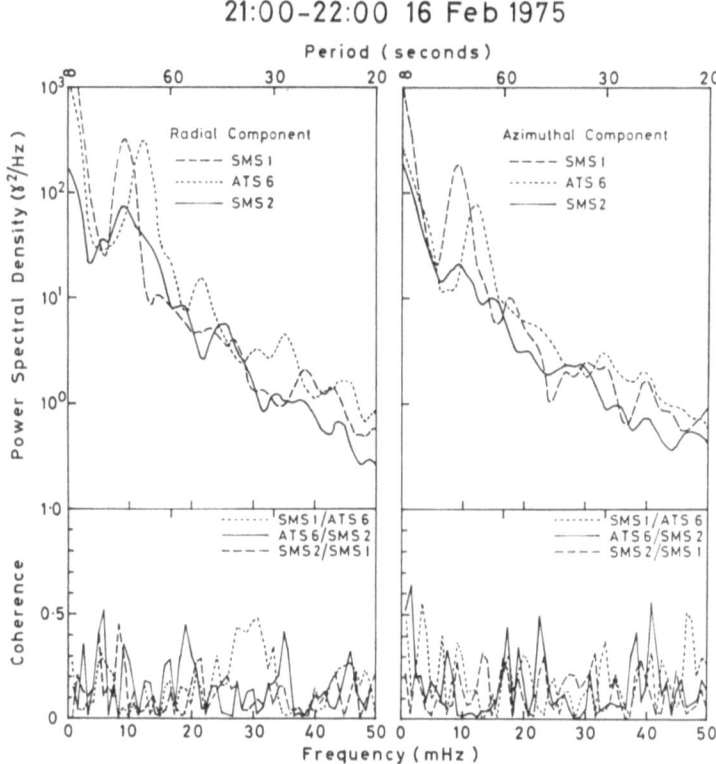

Fig. 9. Spectra of the second hour's data shown in Fig. 8. Note that the coherence is very low (HUGHES *et al.*, 1978a).

was short, about 2,500 km or $m \simeq 100$, there being a whole wavelength between SMS 1 and SMS 2 which were 3.5 degrees apart. The wave had a significant compressional magnetic component and a variety of particle flux oscillations were observed at ATS 6. A sample of these are shown in Fig. 11. Here particle count rate in two representative proton and electron energy channels is plotted against the phase of the magnetic variations. (This procedure was needed as the time for a complete particle energy scan (24 s) was similar to the wave period.) The 1 keV energy protons are modulated by the $E \times B$ drift velocity of the wave. The fact that the oscillation is in quadrature indicates that the wave is a standing oscillation. We can also deduce that the wave is a second harmonic from the phase between E and b and knowing that ATS 6 is above the geomagnetic equator. The higher energy proton flux oscillates in antiphase with b. This is interpreted as due to particle acceleration and deceleration in the wave fields which results in the particle pressure varying in antiphase with the magnetic pressure. $E \times B$ drift velocity is an insignificant perturbation to the 34 eV electrons. Their response is again due to acceleration and deceleration. The electron distribution function indicates that ATS 6 was close to a steep spatial gradient in the 10 keV electron population. Convection of this gradient back and forth past the spacecraft is thought to cause the flux of these electrons to vary in phase with b. The source of this wave was uncertain, but either bounce or drift wave

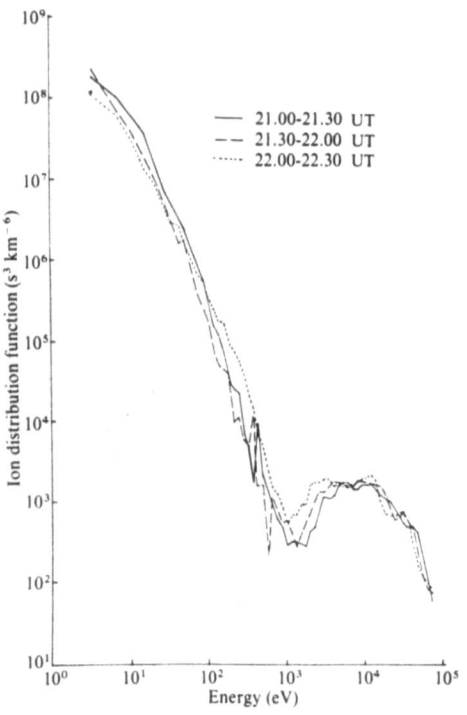

Fig. 10. Ion velocity distribution functions measured on ATS 6 during the three half hour periods during which oscillations are apparent in Fig. 8. The ions have been assumed to be protons (HUGHES *et al.*, 1978b).

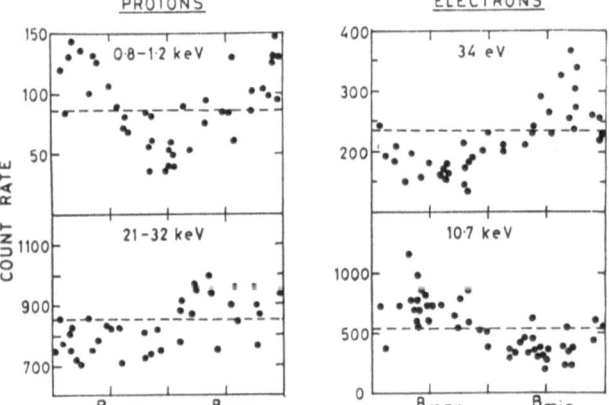

Fig. 11. Oscillations in particle count rates measured on ATS 6 shown versus the phase of the magnetic oscillations occurring at the same time.

plasma instabilities seem the most likely cause.

Pi 2's are the most common type of nighttime pulsation, and they are frequently observed near geostationary orbit (ARTHUR and McPHERRON, 1973; LIN and CAHILL, 1975; McPHERRON, 1979) though interest in them has been limited. Figure 12 shows a Pi 2 pulsation observed near midnight by three geostationary spacecraft. The space-

1975 MARCH 1

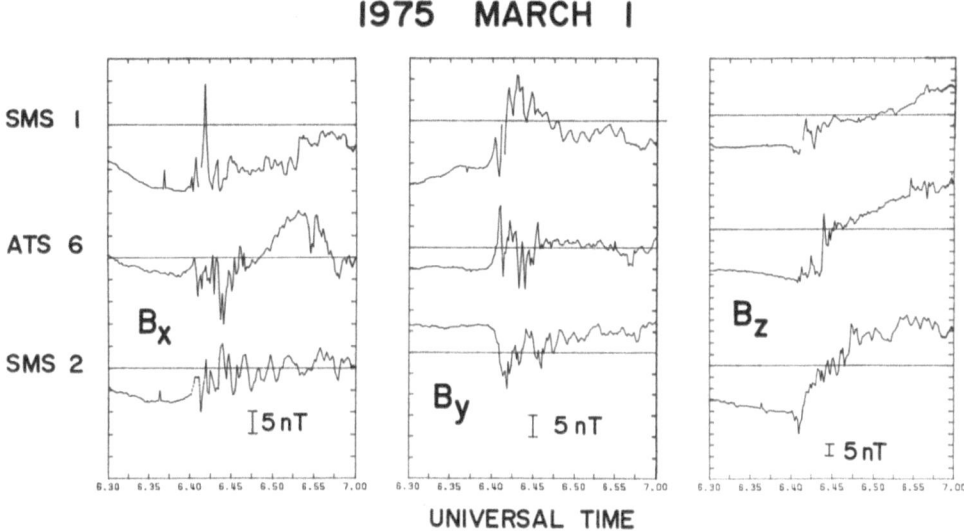

Fig. 12. A Pi 2 pulsation observed by three geostationary spacecraft near local midnight. The data is unfiltered. Spacecraft positions are SMS 1, 75°W; ATS 6, 95°W; SMS 2, 115°W. B_y is measured positive westward.

1975 MARCH 5

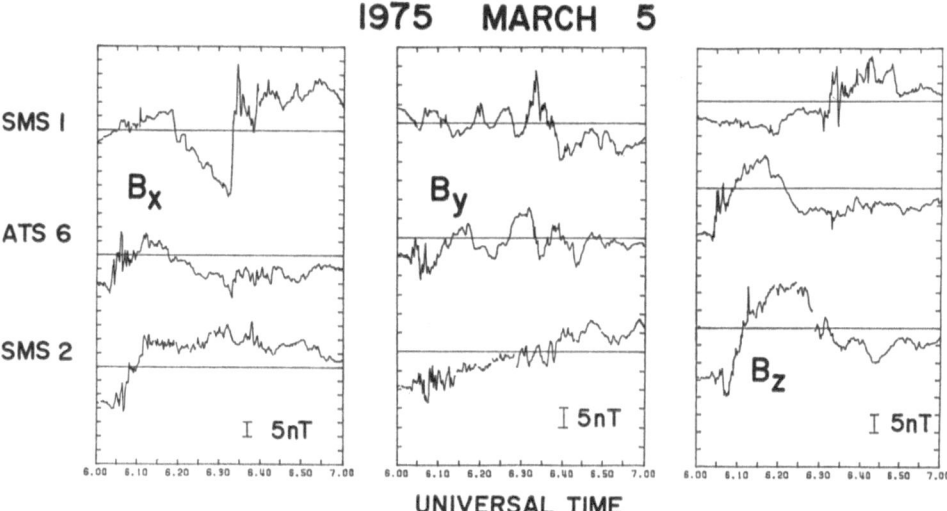

Fig. 13. Similar to Fig. 12 except the first Pi 2 at 6:05 UT is observed only by ATS 6 and SMS 2 while the second at 6:35 UT is seen only by SMS 1.

craft positions are: SMS 1, 75°W; ATS 6, 95°W and SMS 1, 115°W so that there is about a 20° gap between each pair. Local time at ATS 6 is about 0030. The Pi 2 event starts simultaneously at all three spacecraft to within a timing accuracy of a few seconds and is accompanied by a change in the ambient magnetic field which becomes more dipolar. (As B_x is measured radially outward and B_z northwards and as the spacecraft are above the geomagnetic equator, an increase in both B_x and B_z indicates a field dipolarization.) However there is a difference in the DC variation in B_y. SMS 1, located after local midnight, sees a positive, or westward bay; SMS 2, located before local midnight sees a negative eastward variation. This is in keeping with previous statistical studies using a single spacecraft (McPherron, 1979) and is consistent with field aligned currents flowing into the ionosphere on the dawn side of midnight and out of the ionosphere on the dusk side provided they flow on an L shell greater than the spacecraft L shell.

However, Pi 2's are not always seen so globablly at geostationary orbit. From inspecting three weeks data we found that about a third of the events seen were seen at the three spacecraft. Other typical events are shown in Fig. 13. Pi 2's occurred at about 6: 05 and 6: 35 UT when again ATS 6 was very close to local midnight. The first of these events was seen by ATS 6 and SMS 2 but not at SMS 1, the most easterly spacecraft. However the second event was seen only at SMS 1. As in the previous example, the field relaxes to a more dipolar configuration each time a Pi 2 occurs. The field first relaxes at ATS 6 and SMS 1 and than half an hour later at SMS 1. Thus we find that Pi 2 signals can be quite localized at geostationary orbit, but that when they are seen over larger local time sectors, the variation of the signal with local time is consistant with statistical studies.

6. Conclusions

We have tried to demonstrate the power and scope of using multiple satellite data to study ULF waves in the magnetosphere. The technique is clearly important for it enables us to measure parameters such as east/west wavelengths and the scale lengths of resonance regions which it is otherwize hard or impossible to do. Our ability to observe the spatial structure of ULF signals in the magnetosphere allows us to obtain a much better understanding of pulsations.

Significant results to date are the narrowness of resonance regions, which appear to be on the order of $1/2\ R_E$, the extreme difference between morning and afternoon signals in the Pc 4 period band and the localized nature of Pi 2's at geostationary orbit. We expect a lot more work to be done on all these problems using multiple satellite data sets, especially data from the ISEE spacecraft and also geostationary data, during the next few years, and that such work will really stimulate pulsations research.

Figures 3 and 4 were supplied by Dr. H. J. Singer. Discussions with Dr. D. J. Southwood during the original analysis of much of the data are gratefully acknowledged. Support to attend the IUGG General Assembly and present this paper was provided by the NSF (grant # ATM 7911899) and Boston University.

REFERENCES

Allan, W. and F. B. Knox, A dipole field model for axisymmetric Alfvén waves with finite ionospheric conductivities, *Planet. Space Sci.*, **27**, 79–85, 1979.

Arthur, C. W. and R. L. McPherron, Simultaneous ground-satellite observations of substorm-associated Pi 2 micropulsations, *EOS, Trans. Am. Geophys. Union*, **54**, 1175, 1973 (abstract only).

Arthur, C. W., R. L. McPherron, and W. J. Hughes, A statistical study of Pc 3 magnetic pulsations at synchronous orbit, ATS 6, *J. Geophys. Res.*, **82**, 1149–1157, 1977.

Chen, L. and A. Hasegawa, A theory of long-period magnetic pulsations, 1, Steady state excitation of field line resonance, *J. Geophys. Res.*, **79**, 1024–1032, 1974.

COWLEY, S. W. H., Energy transport and diffusion, in *Physics of Solar Planetary Environments*, Vol. 2 edited by D. J. Williams, pp. 582–607, Am. Geophys. Union, Washington, 1976.

CUMMINGS, W. D., R. J. O'SULLIVAN, and P. J. COLEMAN, Standing Alfvén waves in the magnetosphere, *J. Geophys. Res.*, **74**, 778–793, 1969.

DUNGEY, J. W., Electrodynamics of the outer atmosphere, Ionos. Res. Lab., PA State Univ., Report 69, 1954.

DUNGEY, J. W., The structure of the exosphere or adventures in velocity space, in *Geophysics, the Earth's Environment*, edited by C. Dewitt, 537 pp., Gordon and Breach, New York, 1963.

FUKUNISHI, H. and L. J. LANZEROTTI, ULF pulsation evidence of the plasmapause 1. Spectral studies of Pc 3 and Pc 4 pulsations near $L=4$, *J. Geophys. Res.*, **79**, 142–158, 1974.

GREEN, C. A., The longitudinal phase variation of midlatitude Pc 3–4 micropulsations, *Planet. Space Sci.*, **24**, 79–85, 1976.

HERRON, T. J., Phase characteristics of geomagnetic micropulsations, *J. Geophys. Res.*, **71**, 871–889, 1966.

HUGHES, W. J. and D. J. SOUTHWOOD, The screening of micropulsation signals by the atmosphere and ionosphere, *J. Geophys. Res.*, **81**, 3234–3240, 1976.

HUGHES, W. J., R. L. MCPHERRON, and C. T. RUSSELL, Multiple satellite observations of pulsation resonance structure in the magnetosphere, *J. Geophys. Res.*, **82**, 492–498, 1977.

HUGHES, W. J., R. L. MCPHERRON, and J. N. BARFIELD, Geomagnetic pulsations observed simultaneously on three geostationary satellites, *J. Geophys. Res.*, **83**, 1109–1114, 1978a.

HUGHES, W. J., D. J. SOUTHWOOD, B. MAUK, R. L. MCPHERRON, and J. N. BARFIELD, Alfvén waves generated by an inverted plasma distribution, *Nature*, **275**, 43–45, 1978b.

HUGHES, W. J., R. L. MCPHERRON, J. N. BARFIELD, and B. H. MAUK, A compressional Pc 4 pulsation observed by three satellites in geostationary orbit near local midnight, *Planet. Space Sci.*, **27**, 821–840, 1979.

LIN, C. C. and L. J. CAHILL, Pi 2 pulsations in the magnetosphere, *Planet. Space Sci.*, **23**, 693–711, 1975.

MCPHERRON, R. L., Substorm associated micropulsations at synchronous orbit, *J. Geomag. Geolectr.*, **32** Suppl. II, 1980 (this issue).

MIER-JEDRZEJOWICZ, C. A. W. and D. J. SOUTHWOOD, The east/west structure of pulsation activity in the 8–20 mHz band, *Planet. Space Sci.*, **27**, 617–630, 1979.

NEWTON, R. S., D. J. SOUTHWOOD, and W. J. HUGHES, Damping of geomagnetic pulsations by the ionosphere, *Planet. Space Sci.*, **26**, 201–209, 1978.

OLSON, J. V. and G. ROSTOKER, Longitudinal Phase variations of Pc 4–5 micropulsations, *J. Geophys. Res.*, 2481–2488, 1978.

SAMSON, J. C., J. A. JACOBS, and G. ROSTOKER, Latitude dependent characteristics of long-period geomagnetic micropulsations, *J. Geophys. Res.*, **76**, 3675–3683, 1971.

SINGER, H. J., C. T. RUSSELL, M. G. KIVELSON, T. A. FRITZ, and W. LENNARTSSON, Satellite observations of the spatial extent and structure of Pc 3, 4, 5, pulsations near the magnetospheric equator, *Geophys. Res. Lett.*, **6**, 889–892, 1979.

SOUTHWOOD, D. J., Some features of field line resonances in the magnetosphere, *Planet. Space Sci.*, **22**, 483–491, 1974.

SOUTHWOOD, D. J., Comments on field line resonance and micropulsations, *Geophys. J. R. Astr. Soc.*, **41**, 425–431, 1975.

SOUTHWOOD, D. J., J. W. DUNGEY, and R. G. ETHERINGTON, Bounce resonant interaction between pulsations and trapped particles, *Planet. Space Sci.*, **17**, 349–361, 1969.

WALKER, A. D. M., R. A. GREENWALD, W. F. STUART, and C. A. GREEN, Stare auroral radar observations of Pc 5 geomagnetic pulsations, *J. Geophys. Res.*, **84**, 3373–3388, 1979.

Substorm Associated Micropulsations at Synchronous Orbit

R. L. McPHERRON

Department of Earth and Space Science, and Institute of Geophysics and Planetary Physics, University of California, Los Angeles, California, U.S.A.

(Received June 28, 1980)

Magnetometers carried by spacecraft in synchronous orbit show that the magnetic field is almost continuously agitated by the presence of hydromagnetic waves. A variety of waves have been identified, several of which are always associated with certain phases of magnetospheric substorms. These types include Pi 2 bursts and Pi pulsations near midnight and IPDP and mixed mode Pc 4–5 waves near dusk. In this report we describe the characteristics of these waves and discuss possible directions for further work.

1. Introduction

The magnetic field at synchronous orbit is almost continuously disturbed by pulsations of period 1–1,000 sec. These pulsations have magnitudes from fractions of a gamma to as much as 3 gamma. Some of these pulsations are associated with magnetospheric substorms, occurring systematically in certain local time sectors and phases of the substorm. Other types occur while substorms are in progress but are probably not caused by the substorm. The purpose of this paper is to review the current status of observations of substorm associated waves at synchronous orbit and suggest areas for further research.

A summary of pulsation occurrence based on ground observations in the auroral zone is given in Fig. 1. This work (McPHERRON *et al.*, 1968) provided the motivation for our synchronous orbit studies. In a dipole field, or in more realistic field models, field lines from the auroral zone pass through, or near, synchronous orbit. Because of this it is reasonable to expect that these same waves might be seen in space.

As indicated in Fig. 1, four types of pulsations are commonly observed in the night side auroral zone during substorms. These include IPDP type of Pc 1 pulsations and mixed mode Pc 4–5 near dusk, Pi pulsations and Pi 2 bursts near midnight, and Pi (c) in the early morning hours. The relation of these four types of pulsations to substorm phases is summarized in Fig. 2. Bursts of Pi 2 are substorm precursors in the sense that they are frequently observed prior to major substorm expansion onsets. Compressional Pi pulsations and Pi (c) occur throughout the expansion phase and early recovery phase. Mixed mode Pc 4–5 and IPDP pulsations occur during the expansion phase.

In following sections of this paper I present a brief summary of our efforts to observe these waves at synchronous orbit. To illustrate the wave characteristics I use data taken from the UCLA fluxgate magnetometers on ATS 1 and ATS 6.

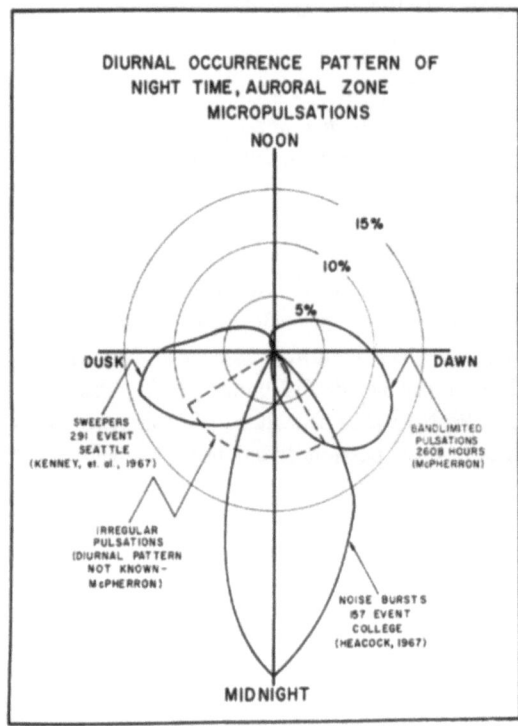

Fig. 1. Diurnal occurrence pattern for substorm associated magnetic pulsations as observed on the ground in the auroral zone.

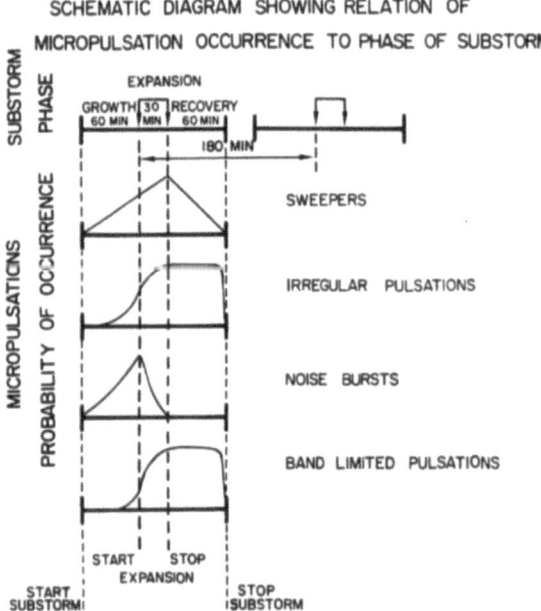

Fig. 2. Relation of magnetic pulsation occurrence observed on the ground in the auroral zone to major substorm onsets.

2. Mixed Mode Pc 4–5 Waves

Mixed mode Pc 4–5 waves are the largest amplitude magnetic pulsations observed at synchronous orbit. Their characteristics as observed near the magnetic equatorial plane by ATS 1 have been extensively documented in a series of papers by Barfield and collaborators (BARFIELD and COLEMAN, 1970; BARFIELD et al., 1972; BARFIELD and MC-PHERRON, 1972a, b, 1978). Additional properties have been reported by LANZEROTTI et al. (1975). An example of an event recorded by the ATS 6 spacecraft is presented in Fig. 3.

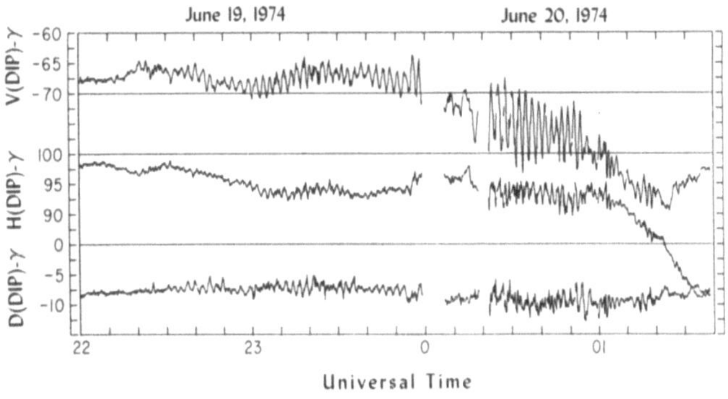

Pc4 MAGNETIC PULSATIONS
UCLA Fluxgate Magnetometer ATS-6

Fig. 3. Mixed mode Pc 4–5 magnetic pulsations observed at synchronous orbit by the ATS 6 spacecraft (H antiparallel to dipole, D azimuthal east transverse to magnetic meridian, V approximately radial outward).

Mixed mode Pc 4–5 waves have a quasi-sinusoidal waveform with amplitudes from 10 to 30 gamma and periods of 50 to 200 sec. The name "mixed mode" indicates that the waves typically have both transverse and compressional components, usually confined to the magnetic meridian plane. Spectra of these waves (Fig. 4) usually have a very narrow band of excited frequencies. Often harmonics of the lowest frequency appear to be present. Dynamic spectra do not show any significant variations of signal frequency with time. Wave polarization is usually linear as shown schematically in Fig. 5, and at a large angle to the ambient field. The properties of the waves are apparently highly localized in space. Although they occur simultaneously over a large local time sector (Fig. 6) spectral analysis shows slightly different frequencies are present at different locations, and the signals are incoherent between spacecraft only 5 degrees apart (HUGHES et al., 1978a). At such times the proton distribution has been found to be peaked at about 10 keV (HUGHES et al., 1978b, see also accompanying paper by HUGHES, 1980).

Mixed mode Pc 4–5 waves occur in the afternoon-dusk sector of local time. They are most often seen, and largest in amplitude during intervals of high magnetic activity. The onset of these pulsations is closely associated with the onset of a substorm expansion near midnight. They usually accompany a decrease in the H component (parallel to dipole axis) at the spacecraft, and occur in the region of the partial ring current development.

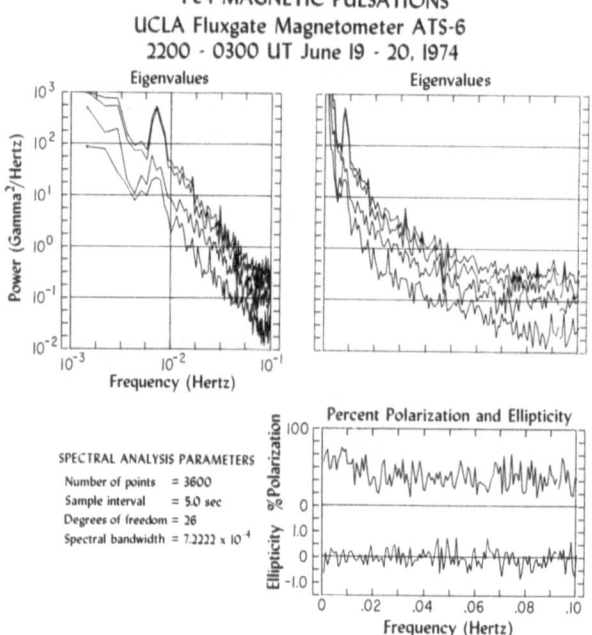

Fig. 4.   Cross spectrum of mixed mode Pc 4–5 wave event shown in Fig. 3.   Note the spectral
matrix has been transformed to the principal axis coordinate system at each frequency.

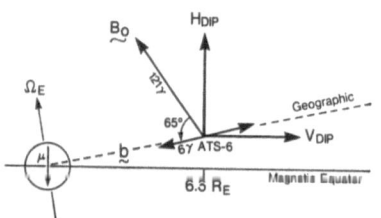

Polarization of Pc4 Magnetic Pulsations
UCLA Fluxgate Magnetometer ATS-6
2230-0215 UT 19 & 20 June 1974

Fig. 5.   Polarization of mixed mode Pc 4–5
signal displayed in Figs. 3 and 4. The wave
is linearly polarized in a meridian plane with
a substantial compressional component.

They are often associated with the occurrence of IPDP pulsations, both at the spacecraft
and the ground (MALTZEVA et al., 1980).   The amplitude of the IPDP sometimes appear
to be modulated by the Pc 4–5 waves as illustrated in Fig. 7.   Modulation of energetic
particles by these waves is frequently observed (SU et al., 1977).

    A possible generation mechanism for mixed mode Pc 4–5 waves is bounce resonance
with ring current protons (SOUTHWOOD, 1969).   Although no detailed theory is presently
available, the observation of HUGHES et al. (1978b) suggests that the source of free energy
for this interaction is the 10 keV peak in the proton distribution.   Such a peak is a natural
consequence of magnetospheric convection as shown by COWLEY (1976).   The appearance
of these waves at synchronous orbit may simply be a result of inward motion of convection

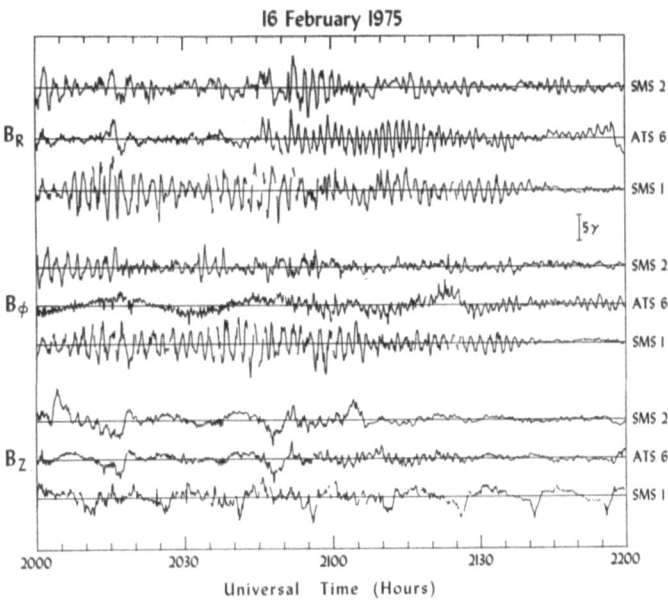

Fig. 6. Pc 4 magnetic pulsations observed simultaneously at three synchronous spacecraft in the late afternoon. Wide spread wave activity is incoherent between spacecraft.

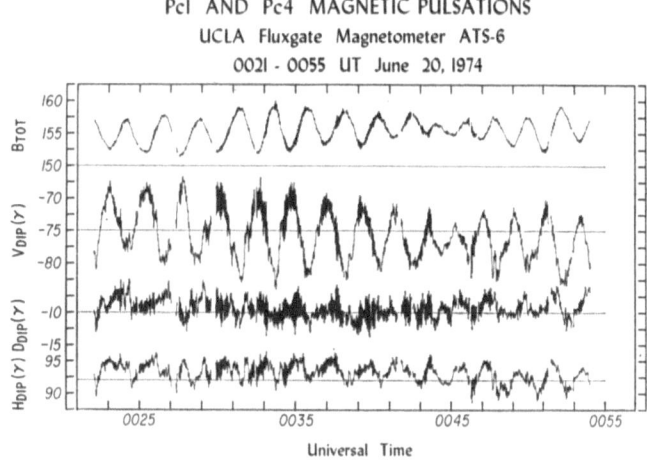

Fig. 7. Apparent modulation of high frequency Pc 1 waves by the mixed mode Pc 4–5 wave shown in Figs. 3–5.

boundaries during substorms. In support of this conjecture (HEDGECOCK, 1976) finds that these waves are present at locations outside synchronous orbit, even during quiet conditions. Important experimental questions concerning these waves include their radial distribution in space, the plasma characteristics required to generate them, their nodal structure along field lines, and the mechanisms for modulating both ambient particles and high frequency waves.

3. Pc 1 Magnetic Pulsations

Magnetic pulsations in the Pc 1–2 frequency band are a common occurrence at synchronous orbit.   Their characteristics have been studied in considerable detail by several authors using data from ATS 1, ATS 6, and GEOS 1 (Barfield and McPherron, 1972; Bossen *et al.*, 1976; Gendrin *et al.*, 1978).   A typical example of the waveform of these waves as observed at ATS 6 is shown in Fig. 7.   The signal consists of nearly sinusoidal oscillations of the field, amplitude modulated in the form of wave packets.   The perturbations are predominantly transverse to the ambient field with amplitude of order 5 gamma peak-to-peak and period 5 sec.   Events typically last about 30 min.   The spectrum of the waves (Fig. 8) shows a broad band of excited frequencies.   The polarization is always left hand elliptical with ellipticities of order 0.5.   The direction of propagation is approximately along the ambient field.   Dynamic spectral analysis (Fig. 9) shows the wave events are made up of a number of narrow band bursts of slightly different frequencies. Normally, the wave frequency tracks a fraction, 0.1 to 0.2, of the local proton gyro frequency (Fig. 10).

Pc 1 activity at synchronous orbit is observed on about every third day.   It is most probable in the afternoon to dusk local time sector.   Pc 1 events are usually associated with major substorm expansion onsets, occurring within the first one and a half hours after the onset Pc 1 events which occur at other times are periodically structured, weaker and associated with very quiet conditions, i.e. pearl type Pc 1.   It has been reported that

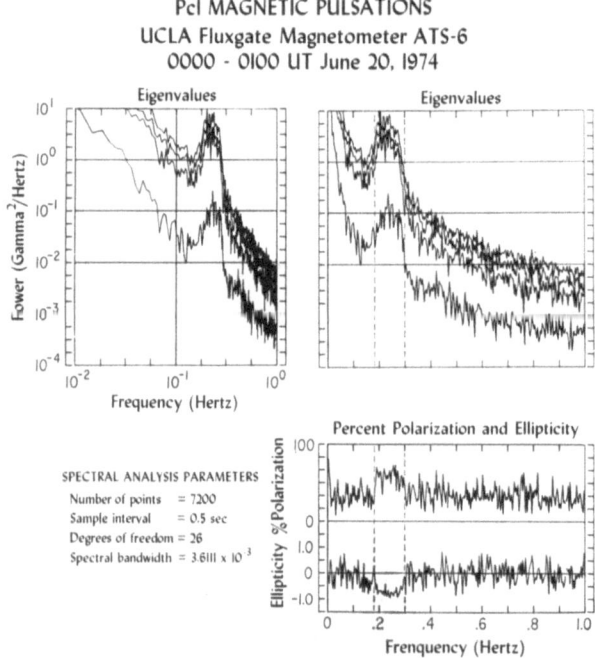

Fig. 8.   Spectral analysis of Pc 1 event of Fig. 7 showing a high degree of left elliptical polarization transverse to ambient field.

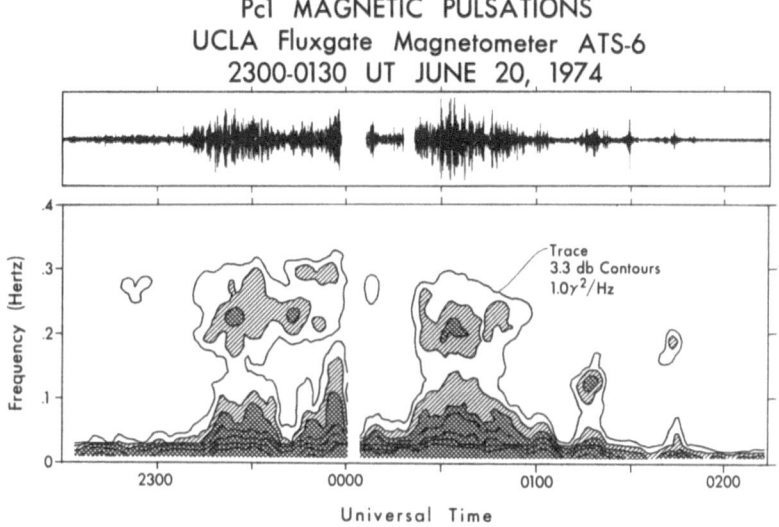

Fig. 9.   Dynamic spectrum of Pc 1 magnetic pulsation event shown in Figs. 7 and 8.

Pc 1 events are related to solar wind dynamic pressure enhancements (KAYE and KIVELSON, 1979).   The author, however, finds that ATS 6 events are statistically associated with solar wind stream boundaries, which are known to be responsible for creation of the southward interplanetary magnetic fields that cause substorms (ROSENBERG and COLEMAN, 1978).

Synchronous orbit Pc 1 activity is correlated with simultaneous Pc 1 activity on the ground (McPHERRON et al., 1972; BOSSEN et al., 1976; GENDRIN et al., 1978).   A typical event is illustrated in Fig. 10.   The usual ground signature is an upward sweep in frequency called an IPDP type of Pc 1.   The satellite signature, however, tracks the local proton gyro frequency.

As discussed earlier in reference to Fig. 7, Pc 1 activity is often associated with mixed mode Pc 4–5 waves (BARFIELD and COLEMAN, 1970; MALTZEVA et al., 1980).   Also, the H component of the field at the satellite often decreases at times of Pc 1 activity suggesting that the satellite has been engulfed in a high beta plasma (BARFIELD and McPHERRON, 1972).   This suggestion is supported by the KAYE and KIVELSON's (1979) report that most of the Pc 1 events seen outside synchronous orbit are correlated with enhanced cold plasma density.   MAUK and McPHERRON (1980) find that events are observed when cold plasma appears in addition to hot ring current protons.   Mauk and McPherron also show that Pc 1 waves observed at ATS 6, 10 degrees above the magnetic equator, are always propagating away from the equator.   GENDRIN et al. (1978) find that the time to propagate from synchronous orbit to the ground is of order 60–70 sec.

Pc 1 magnetic pulsations are almost certainly generated by the ion cyclotron instability of ring current protons.   A possible sequence of events leading to the production of IPDP may be the following.   An enhancement of magnetospheric convection leads to inward drift, energization, and the growth of pitch angle anistropy of ring current protons.   Simultaneously, cold plasma in the dusk sector is stripped off the plasmasphere

SIMULTANEOUS Pc1 EVENT MARCH 18, 1967

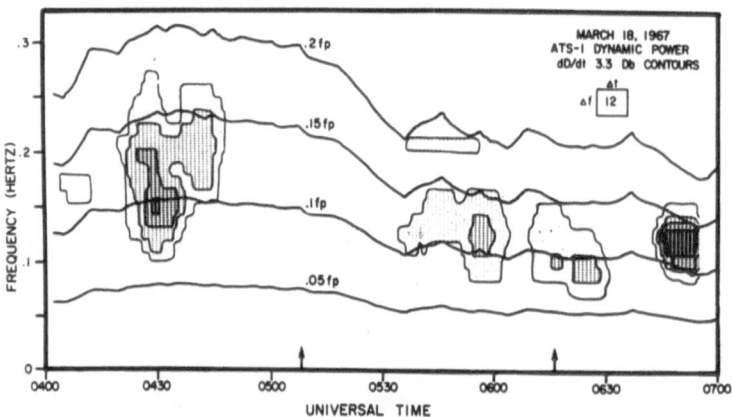

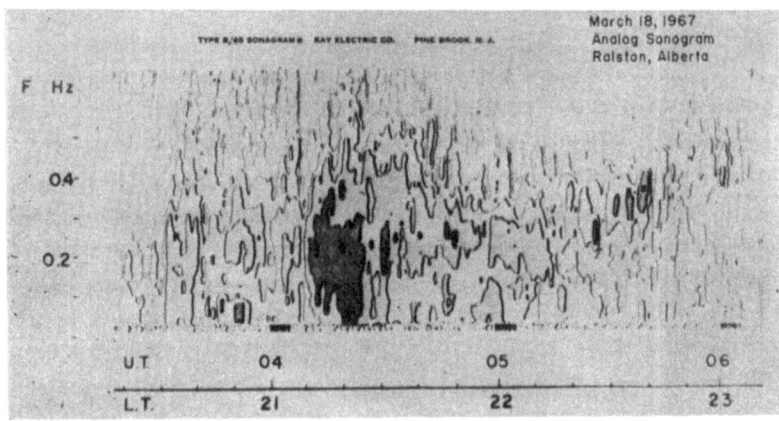

Fig. 10.   Observation of IPDP event on ground during Pc 1 event at synchronous spacecraft
ATS 1.   Wave frequency at satellite tracks local proton gyrofrequency while frequency
at ground sweeps upward.

and begins to drift sunward.   A substorm expansion in the midnight sector causes further
energization of a discrete cloud of protons which drift westward through the regions of
detached cold plasma.   As they pass through this region they become unstable to the
growth of ion cyclotron waves.   The waves scatter protons into the loss cone and they
propagate out of the equatorial plane toward the auroral ionosphere.   At the ionosphere
they enter the ionospheric wave guide and propagate horizontally for great distances.
Leakage from the waveguide is observed on the ground as pulsations.   An upward sweep
in frequency may be caused by two effects, drift dispersion of protons in the equatorial
plane which allows high energy protons (resonant with lowest frequency waves) to arrive
first at the longitude of a station, and by a general inward drift of the entire cloud of
energetic protons.   The absence of an upward sweep of frequency at the satellite is
presumably due to the fact that only signals propagating along the field line through the
satellite can be observed and because the local gyrofrequency is more important in deter-

mining wave frequency than the energy of drifting protons.

There are several interesting questions concerning substorm associated Pc 1 magnetic pulsations which remain to be answered. First, we do not yet have any detailed explanation for the upward sweep of frequency seen at the ground. As mentioned above, this may be due to either inward drift or azimuthal dispersion of clouds of resonant protons. Simultaneous observations at several spacecraft separated azimuthally and/or radially are needed to answer this question. Second, we do not know what effects heavy ions will have on the dispersion relation for Pc 1 propagation. If heavy ions are present there should be observable effects on the spectrum such as switches in polarization with frequency and absorption bands. Combined plasma and field observations are needed for these studies. Third, there is the question of the importance of field aligned ducts in guiding the waves along field lines. Radial profiles of plasma data in conjunction with wave measurements are needed for these studies.

4. Pi 2 Pulsation Bursts

Pi 2 pulsation bursts are seen at synchronous orbit during almost every substorm which occurs as the satellite passes through the midnight sector. Numerous studies of these waves using ground data have been made as discussed in the review by SOUTHWOOD and STUART (1979). To date, however, there are only a few reports of observations of these waves in space (LIN and CAHILL, 1975; COLEMAN and McPHERRON, 1976; ARTHUR and McPHERRON, 1980a). Because of the close association of these bursts with substorm onsets there is a growing interest in their possible role in substorm triggering. A typical example of a Pi 2 burst observed at the synchronous spacecraft ATS 6 is presented in Fig. 11. The bursts have a very irregular waveform, beginning suddenly and fading to background in 5 to 15 min. Most bursts have a strong compressional component and often have a high frequency rider superimposed. Occasionally it is possible to discern the same periodicity at the spacecraft as seen simultaneously at the ground. Typical values are 100 seconds. In a recent study of a number of events at ATS 6, SAKURAI and McPHERRON (1980) find that there are three distinct classes of Pi 2 bursts based on waveform. The first class for which an example is shown in Fig. 11, has a compressional component and a high frequency rider. The second class is similar except that no high frequencies are present. The third class consists of a wave with relatively pure frequency and azimuthal polarization transverse to the main field as illustrated in Fig. 12.

Spectral analysis of individual Pi 2 bursts using normal methods does not appear to be particularly useful because the short duration of the events does not provide sufficient information to calculate accurate spectra. None-the-less SAKURAI and McPHERRON (1980) have used low resolution, Fast Fourier transform methods to show the most typical burst period is about 100 sec, although many events appear to be made up of several harmonics. ARTHUR and McPHERRON (1980b) using spectra of two hour intervals show that azimuthally polarized waves of 100 second period are quite common near midnight. In a separate report, ARTHUR and McPHERRON (1980a) show that the high frequency rider is simultaneously present at synchronous orbit and the ground.

Determination of the wave polarization is more difficult than wave frequency since

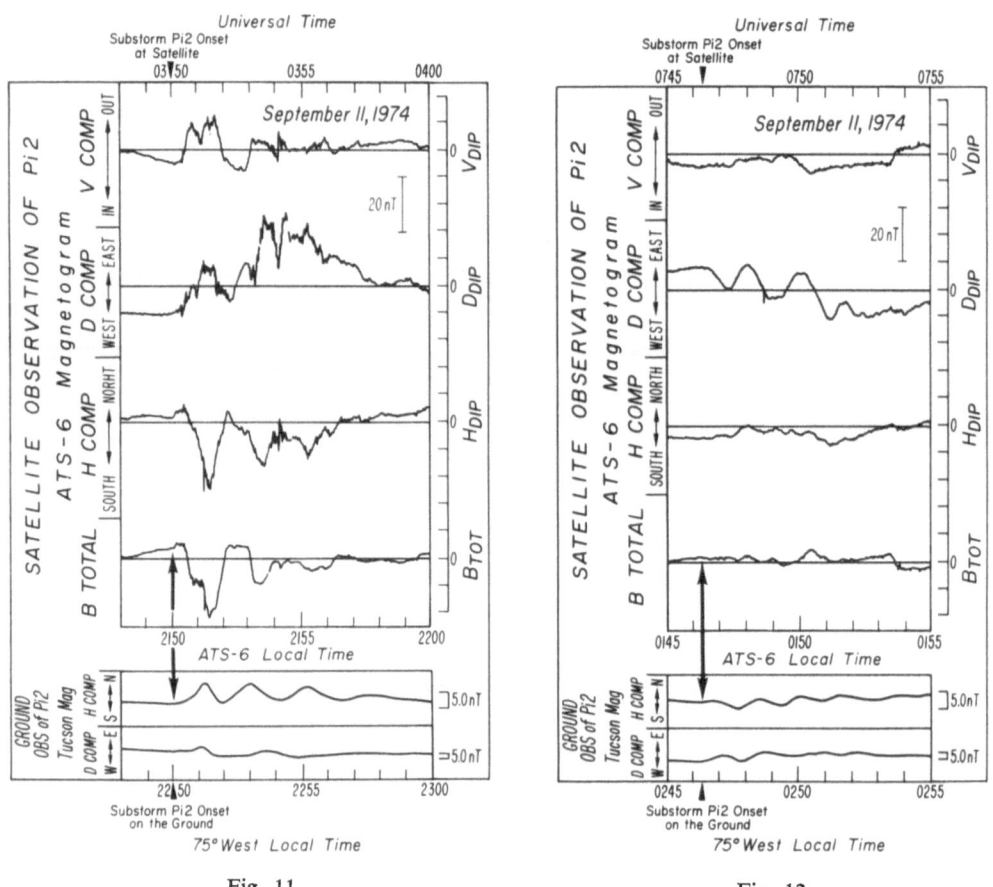

Fig. 11   Fig. 12

Fig. 11. A compressional Pi pulsation burst observed at ATS 6 and simultaneously at
a midlatitude ground station. This burst is of the first type and has a compressional
component and a high frequency rider. Bursts of the second type have no rider.

Fig. 12. The third type of Pi 2 burst seen at ATS 6. The wave is almost completely azimuthal polarized and transverse to main field.

this parameter is even more affected by short records. Examination of wave forms such
as those shown in Fig. 13 leads one to the conclusion that the initial deflection in the
azimuthal component of a Pi 2 burst is positive premidnight and negative post midnight.
This is in the same sense as observed simultaneously in the perturbations due to field
aligned currents.

Pi 2 bursts are observed throughout the night sector with greatest probability about
2200 local time. Virtually every Pi 2 burst seen at the satellite can also be seen at ground
stations in the same meridian as the satellite. The bursts are always associated with
onsets of substorm expansions. In multiple onset substorms a Pi 2 burst accompanies
each onset. As discussed above the bursts are closely correlated with changes in field
aligned currents.

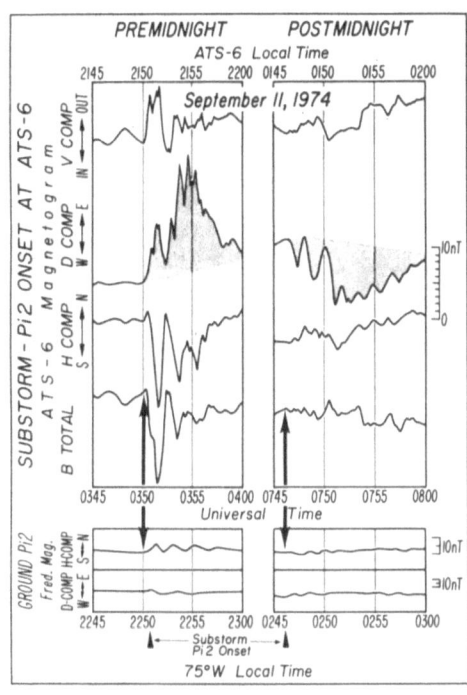

Fig. 13. Pi 2 bursts at ATS 6 are closely associated with magnetic perturbations caused by field aligned currents. The initial deflection in the azimuthal component of the burst is in the same direction as that due to the field aligned current.

The generation mechanism for Pi 2 pulsation bursts is not known. Current speculation suggests that it is a transient associated with a short circuit of the tail current to the auroral oval via field lines. It has been suggested that these waves play an essential role in communicating stress between the magnetosphere and the ionosphere at the onset of rapid convection (SOUTHWOOD and STUART, 1979). Periodicity of Pi 2 is probably controlled by the length of the field lines on which the waves are generated. The smooth waveform of the pulsations at lower latitudes on the ground is thought to be a consequence of an interaction of the waves with the plasmapause.

At the present time the study of Pi 2 bursts is only beginning and there are many fruitful areas for future experimental and theoretical investigation. Maximum entropy spectral analysis should be used to obtain better frequency resolution. Multiple spacecraft should be used to study propagation transverse to the ambient field. Ground-satellite studies should be performed to study field aligned propagation. Data from eccentric spacecraft should be examined to determine the region in which the Pi 2 bursts originate. Properties of the high frequency riders should be determined.

5. Broad Band Pi Pulsations

Another phenomenon observed in the midnight sector during substorms is broad band Pi pulsations. An example of these pulsations as observed at ATS 1 is presented in Fig. 14. Their distinguishing characteristic is an irregular waveform with fluctuations in all components. The spectrum of these waves normally has no identifying property other than a tendency for the compressional component to dominate (McPHERRON, 1970).

Fig. 14.   An example of broad band Pi pulsations accompanying the expansion phase of
a substorm as observed at ATS 1.

These waves are most probable near midnight and are associated with the expansion
phase of substorms.   There appears to be a progressive delay in their onset relative to that
of the substorm as the satellite is located at earlier local times.

The most likely generation mechanism for these waves is turbulence caused by rapid
convection during the substorm expansion phase.   At the present time very little is known
about these waves as no one has studied them either experimentally of theoretically.

6.   Diurnal Occurrence Pattern for Substorm Pulsations

The diurnal occurrence pattern at ATS 1 for several of the substorm pulsation
types discussed above is summarized in Fig. 15.   Mixed mode Pc 4–5 waves are a late
afternoon phenomenon ocassionally occurring as early as noon and rarely extending
beyond dusk.   The number of occurrences at ATS 1 in one year was about 80, or once

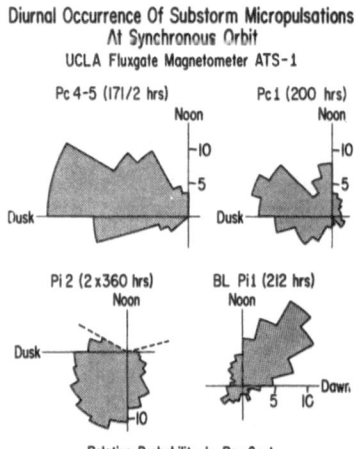

Fig. 15.   Diurnal occurrence pattern for substorm micropulsations at ATS 1. Band limited (BL) pulsations in lower right have been
identified more recently as Pc 3 pulsations
and are probably not substorm related.

every 5 or 6 days.   Transverse Pc 1 waves have a similar, but broader distribution in local
time, and are somewhat more frequent, occurring approximately 100 times per year.
Since the Pc 1 statistics include some quiet time pearl events the occurrence rate is about
the same as for mixed mode Pc 4–5, although there is no one-to-one relation between
the two phenomena.   Broad band Pi 2 pulsations occur in the dusk to midnight sector and
are most probable about 2200 local time.   These pulsations are seen extremely often
with an occurrence rate of about 720 times per year.   Pi 2 bursts were difficult to detect
in the ATS 1 data, but ATS 6 results show they occur throughout the night sector with
greatest probability just before midnight.   The lower right portion of Fig. 15 shows results
for Pc 3 pulsation occurrence at ATS 1 which were at first thought to be substorm related
as discussed in a later section.

7.   Relation of Substorm Micropulsations to Expansion Onset

The relation of the major wave types to expansion onset is presented in Fig. 16.
Mixed mode Pc 4–5 waves generally occur during conditions which are so disturbed it is

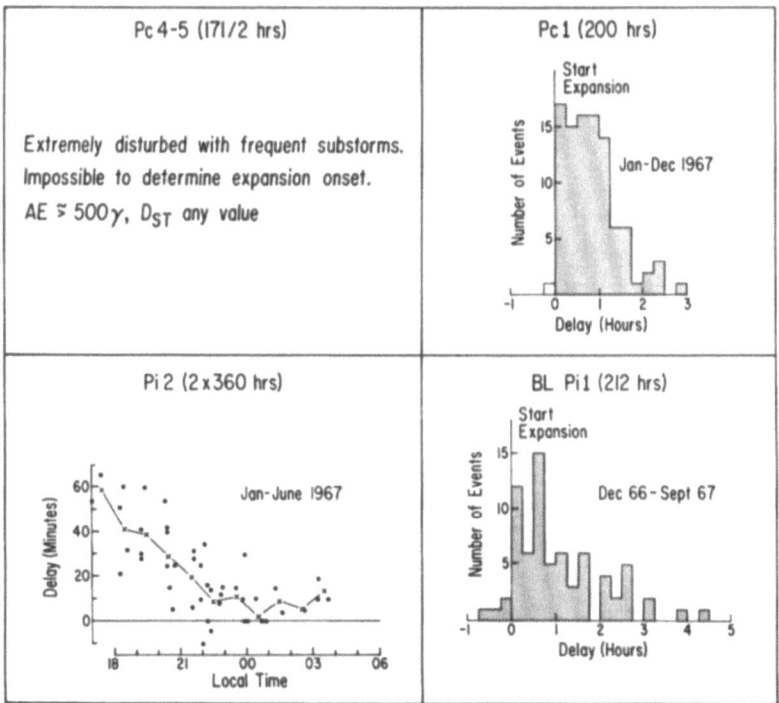

Fig. 16. Relation of various pulsation types to substorm expansion onset at ATS 1. An
onset was defined as the largest, sudden change in auroral zone and midlatitude magnetograms.

difficult to time expansion onset.   During a few, isolated events we can show that the waves begin very soon after expansion onset.   The transverse Pc 1 magnetic pulsations also occur after expansion onset, usually within the first hour.   Broad band Pi 2 are always associated with substorm expansion and recovery phase and show an increasing delay toward earlier local times.   The relation between Pi 2 bursts and substorm onsets is somewhat complicated by the fact that ground Pi 2's are used to define onset.   Using the same definition of substorm onset as for the other wave types, i.e. a large sudden negative bay in the auroral zone and a major positive bay at midlatitudes, we find that a Pi 2 burst is always simultaneous with expansion onset.   However, more than one Pi 2 burst can occur in a single substorm, both before and after a major onset.   Recent changes in terminology would make the first burst the expansion onset and later Pi 2's would be intensifications.

8.   Discussion

Based on the preceding review we conclude that there are four major types of magnetic pulsations associated with magnetospheric substorms.   In our discussion we referred to these as mixed mode Pc 4–5, transverse Pc 1, Pi 2 bursts and broad band Pi pulsations. In addition, we noted that a distinct component of the Pi 2 burst is the high frequency rider which probably has a different origin than the lower frequency component.   When we compare the waves seen in space (summarized in Figs. 15 and 16) to those seen on the ground (Figs. 1 and 2), we note some differences.   First, the mixed mode Pc 4–5 waves are not seen on the ground, and second, the Pi (c) pulsations of the auroral zone are not seen in space.

A possible explanation for the absence of mixed mode Pc 4–5 on the ground can be found in the work of HUGHES and SOUTHWOOD (1976).   These authors show that any wave localized in space such that transverse wavelengths projected on the ionosphere are shorter than the distance from the ionosphere to the ground can not be observed at the ground.   Since these waves are incoherent over very small azimuthal separations in space, it is easy to understand why they might not be observed on the ground.

The absence of Pi (c) pulsations in space is less easy to understand.   An intensive effort has been made to find these waves at synchronous orbit without success (ARTHUR et al., 1973, 1977; ARTHUR and McPHERRON, 1975, 1977a, b).   At first it was thought that they had been identified, McPHERRON and COLEMAN (1971) (see also Figs. 15 and 16), however, the more detailed studies by Arthur and collaborators proved that the waves seen in space are Pc 3 pulsations.   On the ground there is a gradual transition between Pi (c) before dawn and Pc 3 after.   Often Pc 3 are simultaneously present at both locations but Pi (c) are not.   We believe it is likely that Pi (c) are generated by an instability of field aligned currents entering the auroral oval in the morning sector and that these waves only propagate downwards.

It is interesting to note that despite our present convictions that Pc 3 pulsations are not substorm associated, such an association is suggested by the lower right panel of Fig. 15.   This result was obtained by finding the apparent start of the Pc 3 pulsations and the closest preceding substorm onset.   Since during disturbed times substorms can occur every three hours it is possible that chance alone might produce the results shown. This possibility illustrates the difficulty of proving that a substorm causes a particular phenomenon, unless it has a sudden onset closely correlated with the expansion onset.

An important question concerning these pulsations is the extent to which they are related to each other. For example, is a broad band Pi event made up of a sequence of overlapping Pi 2 bursts? Alternatively, is the event a less sinusoidal than normal, mixed mode Pc 4–5 event? As another alternative it is quite possible that the two phenomena are superimposed on broad band Pi pulsations. In scanning waveform data to prepare the occurrence statistics of Figs. 15 and 16, the author at times found it difficult to decide between these possibilities. Clearly, more sophisticated techniques of analysis are required to distinguish between them.

The occurrence statistics, and most of the results reviewed above, were developed from data obtained from ATS 1. This synchronous spacecraft was located at 140 degrees west longitude, nearly at the intersection of the magnetic and geographic equators. This review, however, has been illustrated by data obtained from ATS six at 94 degrees west longitude, approximately ten degrees above the geomagnetic equator. Comparison of data from these two locations shows obvious differences, although, these differences have not been studied extensively. A recent preprint by BAKER *et al.* (1980) makes this point clearly through a statistical study of electron flux pulsations. The cause of these differences is almost certainly the nodal structure of the pulsations as a function of effective magnetic latitude. Future studies of magnetic pulsations must take these differences into account.

It is not surprising that a magnetospheric substorm should be associated with distinct types of micropulsations. The physical processes which occur during a substorm set up a variety of free energy sources which are dissipated through the growth of different types of waves. Magnetospheric convection is particularly important in this regard as it establishes anisotropies in pitch angles, peaks in the energy distribution and spatial gradients.

At the present time detailed theory is available only from Pc 1 micropulsations. Mixed mode Pc 4–5 waves have been examined to some extent, and the remaining waves hardly at all. Current research efforts, both experimental and theoretical, are focused on Pi 2 pulsation bursts because of the possible insight they may provide into processes responsible for substorm triggering. Considerable experimental work on broad band Pi 2 pulsations will be required before any theory can be attempted.

The author would like to acknowledge the many contributions to this work made by his former students and colleagues, C. W. Arthur, J. N. Barfield, M. Bossen, P. J. Coleman, Jr., W. J. Hughes, T. Sakurai. The original work reviewed here has been supported by a variety of contracts and grants including in part NASA NAS 5-11674, NASA NAS 5-11674, NSF DES 75-10678, ONR N00014-75-c-0396. Preparation of this review was supported by ONR N00014-75-c-0396.

REFERENCES

ARTHUR, C. W. and R. L. MCPHERRON, Micropulsations in the morning sector 2. Satellite observations of 10- to 45-second waves at synchronous orbit, ATS 1, *J. Geophys. Res.*, **80**, 4621–4626, 1975.

ARTHUR, C. W. and R. L. McPHERRON, Micropulsations in the morning sector 3. Simultaneous ground-satellite observations of 10- to 45-second waves near $L=6.6$, *J. Geophys. Res.*, **82**, 2859–2866, 1977a.

ARTHUR, C. W. and R. L. McPHERRON, Interplanetary magnetic field conditions associated with synchronous orbit observations of Pc 3 magnetic pulsations, *J. Geophys. Res.*, **82**, 5138–5142, 1977b.

ARTHUR, C. W. and R. L. McPHERRON, Simultaneous ground-satellite observations of Pi 2 magnetic pulsations and their high frequency enhancement, *Planet. Space Sci.*, 1980a (in press).

ARTHUR, C. W. and R. L. McPHERRON, The statistical character of Pc 4 magnetic pulsations at synchronous orbit, *J. Geophys. Res.*, submitted, 1980b.

ARTHUR, C. W., R. L. McPHERRON, and P. J. COLEMAN, Jr., Micropulsations in the morning sector 1. Ground observations of 10- to 45-second waves Tungsten, Northwest Territories, Canada, *J. Geophys. Res.*, **78**, 8180–8192, 1973.

ARTHUR, C. W., R. L. McPHERRON, and W. J. HUGHES, A statistical study of Pc 3 pulsations at synchronous orbit, ATS 6, *J. Geophys. Res.*, **82**, 1149–1157, 1977.

BARFIELD, J. N. and P. J. COLEMAN, Jr., Storm-related wave phenomena at the synchronous, equatorial orbit, *J. Geophys. Res.*, **75**, 1943–1946, 1970.

BARFIELD, J. N. and R. L. McPHERRON, Statistical characteristics of storm-associated Pc 5 micropulsations observed at the synchronous equatorial orbit, *J. Geophys. Res.*, **77**, 4720–4733, 1972a.

BARFIELD, J. N. and R. L. McPHERRON, Investigation of interaction between Pc 1 and 2 and Pc 5 micropulsations at the synchronous orbit during magnetic storms, *J. Geophys. Res.*, **77**, 4707–4719, 1972b.

BARFIELD, J. N. and R. L. McPHERRON, Stormtime Pc 5 micropulsations observed at the synchronous orbit: A possible source, *J. Geophys. Res.*, **83**, 739–743, 1978.

BARFIELD, J. N., R. L. McPHERRON, P. J. COLEMAN, Jr., and D. J. SOUTHWOOD, Storm-associated Pc 5 micropulsation events observed at the synchronous equatorial orbit, *J. Geophys. Res.*, **77**, 143–158, 1972.

BOSSEN, M., R. L. McPHERRON, and C. T. RUSSELL, A statistical study of Pc 1 magnetic pulsations at synchronous orbit, *J. Geophys. Res.*, **81**, 6083–6091, 1976a.

BOSSEN, M., R. L. McPHERRON, and C. T. RUSSELL, Simultaneous Pc 1 observations by the synchronous satellite ATS 1 and ground stations: implications concerning IPDP generation mechanisms, *J. Atmos. Terr. Phys.*, **38**, 1157–1167, 1976b.

COLEMAN, P. J., Jr. and R. L. McPHERRON, Substorm observations of magnetic perturbations and ULF waves at synchronous orbit by ATS 1 and ATS 6, in *The Scientific Satellite Programme during the International Magnetospheric Study*, edited by K. Knott and B. Battrick, pp. 345–366, D. Reidel, Dordrecht, Holland, 1976.

COWLEY, S. W. H., Energy transport and difusion, in *Physics of Solar and Planetary Environment*, Vol. 2, edited by D. J. Williams, 582 pp., Am. Geophys. Union, Washington D. C., 1976.

GENDRIN, R., S. PERRAUT, H. FARGETTON, F. GLANGEAUD, and J. L. LACOUME, ULF waves: conjugated ground-satellite relationships, *Space Sci. Rev.*, **22**, 433–442, 1978.

HEDGECOCK, P. C., Giant Pc 5 pulsations in the outer magnetosphere: A survey HEOS-1 data, *Planet. Space Sci.*, **24**, 921–935, 1976.

HUGHES, W. J., Multisatellite observations of geomagnetic pulsations, *J. Geomag. Geoelectr.*, **32**, Suppl. II, 1980 (this issue).

HUGHES, W. J. and D. J. SOUTHWOOD, The screening of micropulsation signals by the atmosphere and ionosphere, *J. Geophys. Res.*, **81**, 3234, 1976.

HUGHES, W. J., R. L. McPHERRON, and J. N. BARFIELD, Geomagnetic pulsations observed simultaneously on three geostationary satellites, *J. Geophys. Res.*, **83**, 1109–1116, 1978a.

HUGHES, W. J., D. J. SOUTHWOOD, B. MAUK, R. L. McPHERRON, and J. N. BARFIELD, Alfvén waves generated by an inverted plasma distribution, *Nature*, **275**, 43, 1978b.

KAYE, S. M. and M. G. KIVELSON, Observations of Pc 1–2 waves in the outer magnetosphere, *J. Geophys. Res.*, **84**, 4267–4276, 1980.

LANZEROTTI, L. J., G. G. MACLENNAN, H. FUKUNISHI, J. WALKER, and D. J. WILLIAMS, Latitude and longitude dependence of storm time Pc 5 type plasma waves, *J. Geophys. Res.*, **80**, 1014, 1975.

LIN, C. P. and L. J. CAHILL, Jr., Pi 2 micropulsations in the magnetosphere, *Planet. Space Sci.*, **23**, 693,

1975.

MALTZEVA, N., V. A. TROITSKAYA, R. SCHEPETNOV, O. POKHOTELOV, M. GOCHBERG, V. PILIPENKO, R. L. McPHERRON, and J. BARFIELD, Pc 4–Pc 1 magnetic pulsations at synchronous orbit and their relation to pulsations on the ground, 1980 (in preparation).

MAUK, B. and R. L. McPHERRON, An experimental test of the electromagnetic ion cyclotron instability within the earth's magnetosphere, *Phys. Fluids*, 1980 (in press).

McPHERRON, R. L. and P. J. COLEMAN, Jr., Magnetic fluctuations during magnetospheric substorms 1. Expansion phase, *J. Geophys. Res.*, **75**, 3927–3931, 1970.

McPHERRON, R. L. and P. J. COLEMAN, Jr., Satellite observations of band-limited micropulsations during a magnetospheric substorm, *J. Geophys. Res.*, **76**, 3010–3021, 1971.

McPHERRON, R. L., G. K. PARKS, F. V. CORONITI, and S. H. WARD, Studies of the magnetospheric substorms, II. Correlated magnetic micropulsations and electron precipitation occurring during auroral substorms, *J. Geophys. Res.*, **73**, 1697–1713, 1968.

McPHERRON, R. L., C. T. RUSSELL, and P. J. COLEMAN, Jr., Fluctuating magnetic fields in the magnetosphere, *Space Sci. Rev.*, **13**, 411–454, 1972.

PERRAUT, S., R. GENDRIN, P. ROBERT, A. ROUX, and C. DE VILLEDARY, ULF waves observed with magnetic and electric sensors on GEOS 1, *Space Sci. Rev.*, **22**, 347–369, 1978.

ROSENBERG, R. L. and P. J. COLEMAN, Jr., Solar-cycle dependent north-south field configurations observed in solar wind interaction regions, *J. Geophys. Res.*, to be submitted, 1980.

SAKURAI, T. and R. L. McPHERRON, Satellite observations of Pi 2 activity at synchronous orbit, 1980 (in preparation).

SOUTHWOOD, D. J. and F. W. STUART, Pulsations at the substorm onset, in *Dynamics of the Magnetosphere, Astr. Space Sci. Lib. Series* edited by S.-I. Akasofu, 1980.

SOUTHWOOD, D. J., J. W. DUNGEY, and R. G. ETHERINGTON, Bounce resonant interaction between pulsations and trapped particles, *Planet. Space Sci.*, **17**, 349, 1969.

SU, S.-Y., A. KONRADI, and T. A. FRITZ, On propagation direction of ring current proton ULF waves observed by ATS 6 at 6.6 R_e, *J. Geophys. Res.*, **82**, 1859–1868, 1977.

Low Frequency Pulsation Generation by Energetic Particles

D. J. Southwood

Blackett Laboratory, Imperial College, Prince Consort Road, London, England

(Received June 28, 1980)

We review some ideas developed over the past decade about the generation of low frequency waves in the magnetosphere by wave particle resonance. We emphasise the importance of wave symmetry about the equator, the particular type of particle distributions present in the magnetosphere and differences between drift and bounce resonance. Many arguments are independent of the precise wave mode. Later in the paper we concentrate on the problem of Alfvén wave generation by ring current protons. Recent analysis of measurements at synchronous orbit have yielded supporting experimental evidence.

1. Introduction

Steadily pulsation energy sources are being delimited. A whole series of theoretical and observational works (Dungey, 1954; Atkinson and Watanabe, 1966; Southwood, 1968, 1974; Dungey and Southwood, 1970; Chen and Hasegawa, 1974; Lanzerotti *et al.*, 1974; Hughes *et al.*, 1978a; Olson and Rostoker, 1978) have established that the magnetopause is a major source of magnetospheric hydromagnetic wave energy but it is not the only source. Measurements that show that mid-latitude pulsation propagation on the dayside does not fit the diurnal variation that the Kelvin-Helmholtz instability predicts. Herron (1966), Green (1976), and Mier-Jedrzejowicz and Southwood (1979) demonstrate this. Herron's (1966) work found a predominance of propagation towards the sun. Mier-Jedrzejowicz and Southwood (1979) also found some similar tendency. Whether or not such signal behaviour indicates a nightside source for dayside signals, the prime nightside source of pulsations is the impulsive changes associated with the geomagnetic substorm. The Pi 2 is the particular signal which seems to be excited most directly by the substorm (see Southwood and Stuart (1979) for a recent review) but need not be the only type. In this paper our interest is in a third source of pulsation energy, one that is very interesting and not without controversy. Energy can be fed from the hot plasma (the ring current) into plasma waves at pulsation frequencies. The energy transfer is by particles resonating with components of the wave signal, the process that gives rise to the Landau damping phenomenon in a plasma near thermodynamic equilibrium. Under extreme conditions energy can be released from an inverted hot plasma energy distribution but as we show energy inversion is not a necessary requirement. Spatial gradients alone may be adequate.

Recently, what the Imperial College group regarded as very good internal evidence of waves being generated by hot plasma has emerged in analysis of magnetometer data and particle data recorded (by UCLA, UCSD, and NOAA groups) on ATS 6 and other geosynchronous spacecraft. The basic theoretical notions were published some ten years ago (DUNGEY, 1966; SOUTHWOOD et al., 1968, 1969) and developed in succeeding papers (SOUTHWOOD, 1973, 1976, 1977). We aim to review key points from this work here. We first review how energy is fed from particles to low frequency waves and derive some general results independent of precise wave polarisation. In our view wave interactions with hot plasma should be studied in the light of what is known about the hot plasma distribution and origin. Very immediate conclusions can be drawn from the nature of the distribution (due to adiabatic injection), the symmetry of the magnetic field and the standing structure of low frequency signals; the type of wave most favoured for generation by resonance interactions has nodes in transverse electric field at the equator and antinodes in the transverse magnetic field. We then discuss how these circumstances can be deduced from the data.

The theory here is still controversial. Recent papers such as those by KOZHEVNIKOV et al. (1976), HASEGAWA and MIMA (1978), and HASEGAWA (1979) completely ignore some of the points we regard as crucial, agree with us in other respects and emphasise effects we have downplayed. More theory needs to be done but perhaps more importantly, particle data measured in space has yet to be fully exploited for exploring the internal dynamics of low frequency waves seen in space.

2. Resonance

When a low frequency wave is present in the magnetosphere plasma particles of all energies respond in some way to it. The wave fields will alternately accelerate and decelerate particles as well as convecting plasma back and forth in the L shell due to wave induced $E \times B$ and other drifts. Observable effects such as oscillations in particle flux in a spacecraft detector at the wave frequency naturally result.

How particles of a particular energy behave depends strongly on how they "see" the wave. A particle with a bounce frequency much in excess of the wave frequency makes many bounces in a wave cycle and its response at any particular point in space is a function of the wave amplitude distribution over its entire bounce orbit. In contrast a cold particle which does not move far along the field in a wave period shows a response which depends only on the local wave fields. As well as the relative speed of its bounce motion along the field its rate of east-west motion can radically affect how a particle responds. A full story is complicated and we shall not attempt it here but there is one class of particles whose behaviour in the wave is special. These are those particles which bounce and drift through the wave field in such a way as to see a steady component in the wave signal. Such particles are resonating with the wave. The condition for resonance is that the wave frequency seen by a particle, that is the wave frequency Doppler shifted by magnetic ∇B and curvature angular drift $\tilde{\omega}_d$, is a multiple of its bounce frequency, ω_b; i.e.

$$\omega - m\tilde{\omega}_d = N\omega_b \qquad (N, \text{ integer})$$

where m is the east-west angular wave number.

The bulk of the plasma population, the particles that do not resonate with the wave, behave reactively (rather than resistively) in the wave. This means that the bulk of the plasma only strongly affects the real part of the wave frequency. On the other hand the resonant particles exchange energy with the wave and because of this contribute to the wave damping or growth rate. We can examine this energy exchange by computing $\langle j_{\text{res}} \cdot E \rangle$ where j_{res} is the current in the wave produced by the resonant particles. We do this computation in the next section. It is done without making prior assumptions about wave polarisation. The form of the answer shows this is worthwhile. Whether particles feed energy to the wave or vice versa depends on whether a derivative, df/dW, of the resonant distribution function is positive or negative respectively.

$$\frac{df}{dW} = \frac{\partial f}{\partial W} + \frac{m}{qB_{\text{eq}}\omega} \frac{\partial f}{\partial L} .$$

W, L, and B_{eq} are energy, L shell and equatorial magnetic field. It is clear that df/dW may be positive even if $\partial f/\partial W$ is negative. In fact $\partial f/\partial W$ is positive at times in the inner ring current as we point out.

In a dipole magnetic field ω_{d} and ω_{b} vary as v^2 and v (particle velocity) respectively and are not strongly dependent on pitch angle (HAMLIN et al., 1961). For each value of N the resonance condition is a quadratic so each species can provide two sets of resonant particles. The $N=0$ resonance, the drift resonance, is exceptional as there is only one set of resonant particles. Eastward moving waves can resonate with electrons drifting at the wave east-west phase speed whilst westward moving waves can only drift resonate with ions. Another special feature of the $N=0$ resonance is that first and second adiabatic invariants, μ, J, are conserved by the resonance. This is of some importance in our later discussion.

Inspection of df/dW shows that large m and small ω are favoured for instability as this can maximize the effect of the gradient term. Some waves, such as the transverse Alfvén wave that we concentrate on later, have ω independent of m. Conditions for small ω and large m can then be posed independently. Large m separates the roots of the resonance velocity quadratic. One finds for $N=1$, for instance, a high energy resonant group with

$$m\tilde{\omega}_{\text{d}} \simeq \omega_{\text{b}}$$

and a low energy set with

$$\omega \simeq \omega_{\text{b}}$$

(SOUTHWOOD et al., 1969; SOUTHWOOD, 1976). Because the bounce frequency is a function of velocity a wave with m value appropriate to resonate with ring current protons ($m \sim 100$ for 10 keV at $L=6.6$) resonates with a much more energetic electron population, one thus unlikely to be present in significant numbers. Instability in such a case is thus a competition between ring current ions feeding energy in and damping by low energy resonant particles and any other sources of damping present.

3. Transfer of Energy between Particles and Waves

We calculate the exchange of energy between a group of resonant particles and a wave by computing the mean value of $j_{\text{res}} \cdot E$ where j_{res} is the current due to the resonant distribution. In the low frequency ($\omega \ll$ gyrofrequency), long wavelength ($\gg$ Larmor radius) limit, the hydromagnetic limit, the hot plasma current perpendicular to the field is proportional to plasma pressure (see, e.g. CLEMMOW and DOUGHERTY, 1969). Writing the wave pressure perturbations as $\delta p_{\parallel}$, $\delta p_{\perp}$ we have

$$ \boldsymbol{j}_{\perp} = -\left[\nabla(\delta p_{\perp}) + (\delta p_{\parallel} - \delta p_{\perp})\frac{\hat{n}}{R} \right] \times \frac{\boldsymbol{B}}{B^2} \tag{1} $$

where $\hat{n}$ is the field principal normal, R the field curvature radius. The contribution that this makes to the volume integral of $\boldsymbol{j} \cdot \boldsymbol{E}$ can be rearranged in the following way

$$ \int \mathrm{d}^3 r \boldsymbol{j}_{\perp} \cdot \boldsymbol{E} = \int \mathrm{d}^3 r \left[\nabla(\delta p_{\perp}) + (\delta p_{\parallel} - \delta p_{\perp}) \right] \cdot \frac{\boldsymbol{E} \times \boldsymbol{B}}{B^2} $$

$$ = -\int \mathrm{d}^3 r[(\nabla \cdot \boldsymbol{u}_E) + \boldsymbol{u}_E \cdot \hat{n}/R]\delta p_{\perp} + \int \mathrm{d}^3 r \boldsymbol{u}_E \cdot \hat{n}\delta p_{\parallel}/R \tag{2} $$

by Gauss' theorem, where

$$ \boldsymbol{u}_E = \boldsymbol{E} \times \boldsymbol{B}/B^2 . $$

The electric field, $\boldsymbol{E}$, must be related to the wave magnetic perturbation $\boldsymbol{b}$ by Faraday's law, vector manipulation of which (SOUTHWOOD, 1973, 1976) shows

$$ -B(\nabla \cdot \boldsymbol{u}_E + \boldsymbol{u}_E \cdot \hat{n}/R) = \partial b_{\parallel}/\partial t + \boldsymbol{u}_E \cdot \nabla B . $$

Thus,

$$ \int \mathrm{d}^3 r \boldsymbol{j} \cdot \boldsymbol{E} = \int \mathrm{d}^3 r \frac{\delta p_{\perp}}{B} \frac{\partial b_{\parallel}}{\partial t} + \int \mathrm{d}^3 r \boldsymbol{u}_E \cdot \left(\frac{\delta p_{\perp} \nabla B}{B} + \frac{\delta p_{\parallel} \hat{n}}{R} \right) $$

$$ = \int \mathrm{d}^3 r \int \mathrm{d}^3 v \left(\frac{\mu \partial b_{\parallel}}{\partial t} + q\boldsymbol{E} \cdot \boldsymbol{v}_{\mathrm{d}} \right) \delta f \tag{3} $$

where μ is the magnetic moment and $\boldsymbol{v}_{\mathrm{d}}$ is the combined ∇B and curvature magnetic drift. Allowing for parallel current and electric field one has

$$ \int \mathrm{d}^3 r \boldsymbol{j} \cdot \boldsymbol{E} = \int \mathrm{d}^3 r \mathrm{d}^3 v \left(qE_{\parallel}v_{\parallel} + q\boldsymbol{E} \cdot \boldsymbol{v}_{\mathrm{d}} + \frac{\mu \partial b_{\parallel}}{\partial t} \right) \delta f = \int \mathrm{d}^3 r \int \mathrm{d}^3 v \dot{W} \delta f \tag{4} $$

where

$$ \dot{W} = qE_{\parallel}v_{\parallel} + q\boldsymbol{E} \cdot \boldsymbol{v}_{\mathrm{d}} + \mu \partial b_{\parallel}/\partial t $$

is the adiabatic acceleration rate (NORTHROP, 1963) due to the wave (W = particle energy). We can regard the background distribution as a function of constants of motion, μ, W and magnetic shell parameter, L. In an axisymmetric field L can be defined as the radial distance to the field line equator. μ may be conserved if the wave period and wavelength are large but W and L will always be changed. The Liouville theorem shows, if μ is constant,

$$ \delta f = -\delta W \frac{\partial f}{\partial W} - \delta L \frac{\partial f}{\partial L} \tag{5} $$

where

$$\delta W = \int_{-\infty}^{t} dt \, \dot{W} \tag{6}$$

and (in linear wave theory) the integral is taken along the unperturbed trajectory. Now DUNGEY (1966), SOUTHWOOD et al. (1969), and several subsequent papers pointed out that in resonance the δL and δW were proportional so we only need to compute one or the other

$$\frac{\delta W}{\delta L} = \frac{q\omega}{m} B_{eq} L R_E^2 = \frac{dW}{dL} \ , \ \text{say} \tag{7}$$

where B_{eq} is the equatorial field and we measure L in units of Earth radii, R_E. The trajectory along which we integrate (6) consists of bouncing along $\boldsymbol{B}$ and drift across $\boldsymbol{B}$ at fixed L. The periodicity of the bounce motion means any quantity seen by the particle can be expressed as

$$A = \sum_{N=-\infty}^{\infty} A_E \exp iN\theta$$

i.e. as a Fourier series in particle bounce phase, θ. We can allow for the drift in longitude by expressing the acceleration seen by the particle as

$$\dot{W} = \sum_{N=-\infty}^{\infty} \dot{W}_N \exp i(N\omega_b t + m\tilde{\omega}_d t - \omega t) \tag{8}$$

where ω_b, $\tilde{\omega}_d$ are bounce and drift frequencies and m is angular wave number.

For a weakly growing wave the integration (6) gives

$$\delta W = \sum_N \frac{i \dot{W}_N \exp i(N\omega_b + m\tilde{\omega}_d - \omega)t}{\omega - m\tilde{\omega}_d - N\omega_b} \ .$$

For resonant particles one term dominates and

$$\delta W \simeq \tau_N \dot{W}_N \exp i(N\omega_b + m\tilde{\omega}_d - \omega)t \tag{9}$$

where τ_N is the resonant term

$$\tau_N \simeq \frac{i}{\omega - m\tilde{\omega}_d - N\omega_b} \ .$$

We can now take (9), (7), (5), and (4) together to show

$$\int d^3r R(\boldsymbol{j}_{res}^* \cdot \boldsymbol{E}) = -\sum_i \sum_N \int d^3r \int d^3v |\dot{W}_N|^2 \tau_N \frac{df_i}{dW} \tag{10}$$

where

$$\frac{df}{dW} = \frac{\partial f}{\partial W} + \frac{dL}{dW} \frac{\partial f}{\partial L} \ .$$

Summations are taken over species i and different resonances, N.

Equation (10) illustrates the well-known fact that whether a particular group of resonant particles contributes to wave growth or damping depends on the sign of df/dW (SOUTHWOOD et al., 1969) as we discussed earlier. As SOUTHWOOD (1976) shows groups contribute in proportion to their pressure.

4.   Diffusion in Resonance

One can picture the nett effect of a population of waves on particles in resonance as diffusion in energy and L shell.   Dungey (1966) and Southwood et al. (1969) looked at the problem this way and their relation (7) means the diffusion in L and W is coupled and is thus strictly one dimensional.   For ω, m fixed, energy and L diffusion coefficient are evidently related by

$$D_{LL} = \left(\frac{\mathrm{d}L}{\mathrm{d}W}\right)^2 D_{WW} .$$

This point has been missed in some works (e.g. Hasegawa and Mima, 1978).   The diffusion equation can be written (ignoring sinks and sources of particles)

$$\frac{\partial f_{\mathrm{res}}}{\partial t} = \frac{\mathrm{d}}{\mathrm{d}W} |\dot{W}_N|^2 \tau_N \frac{\mathrm{d}f_{\mathrm{res}}}{\mathrm{d}W}$$

where evidently

$$D_{WW} = |\dot{W}_N|^2 \tau_N .$$

In the diffusion picture the energy released or absorbed by the particle distribution is

$$\sum_i \sum_N \int \mathrm{d}^3 r \int \mathrm{d}^3 v \, W \frac{\mathrm{d}}{\mathrm{d}W} \, D_{WW} \frac{\mathrm{d}f}{\mathrm{d}W} .$$

One integration reduces this to the same form as the r.h.s. of (10).

The diffusion picture provides a rationalisation of the resonant term τ_N.   Looking at the diffusion as a random walk as particles experience interactions with different waves in a noise band we can interpret τ_N as the interaction time and thus $\tau_N \sim 1/\Delta\omega$ where $\Delta\omega$ is the bandwidth of the signals seen by the particle.

5.   The Magnetospheric Ring Current Distribution

Let us now look at the type of particle distribution that occurs in the magnetosphere in the ring current region.   Magnetospheric convection is responsible for injecting this plasma from the tail and consequently the plasma has well-defined properties in energy and space (see e.g. Cowley, 1976).   The injection is adiabatic and the invariants, μ, J, are conserved.   As a particle moves in, energy increases at a well-defined rate $(\partial W/\partial L)_{\mu, J}$. The atmospheric loss cone is small at large L and to a first approximation loss is important in defining the distribution mainly near the inner edge.   Figure 1 shows two distribution function contours in the W, L plane.   The outer legs of each contour should have a slope, $(\mathrm{d}W/\mathrm{d}L)_f$, close to but less negative than $(\partial W/\partial L)_{\mu, J}$ where loss is unimportant as the dotted curve illustrating the variation of W with L shows.   The inner edges are determined by the processes that limit inward motion: loss by collision or the competition between convection and rotation about the Earth (see e.g. Cowley and Ashour-Abdalla, 1976). The contour, f_1, is appropriate for a medium proton energy in the evening and afternoon sectors.   Its inner edge has a negative slope because 10–15 keV protons penetrate to low L more effectively than lower energy protons and the proton distribution has an

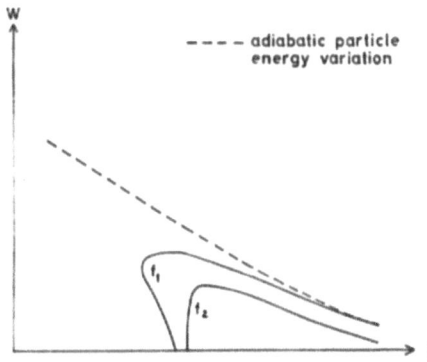

Fig. 1. Schematic representation of distribution function contours for ring current particles in the W, L plane.

energy inversion. (This occurs because corotation and magnetic drifts tend to cancel for such particles; the phenomenon is well-documented, CHEN, 1970; SMITH and HOFFMAN, 1974; MAUK and McILWAIN, 1974; KIVELSON and SOUTHWOOD, 1975; COWLEY and ASHOUR-ABDALLA, 1976; COWLEY, 1976.). Everywhere $(dW/dL)_f > (\partial W/\partial L)_{\mu,J}$.

$(\partial W/\partial L)_{\mu,J}$ is directly related to the particle mean magnetic drift, $\bar{\omega}_d$ (see e.g. NORTHROP and TELLER, 1960)

$$\frac{\partial W}{\partial L}\bigg|_{\mu,J} = -\frac{q}{c}\,\bar{\omega}_d B_{eq} L R_E$$

(taking $\bar{\omega}_d$ positive for ions). Comparing this with (7) shows that particles in the $N=0$ resonance where $\omega = m\bar{\omega}_d$ conserve μ, J in resonance. Early work pointed out that as

$$\frac{dW}{dL}\bigg|_f \gtrsim \frac{\partial W}{\partial L}\bigg|_{\mu,J} \quad \text{and thus } (\partial f/\partial L)^{\mu,J} \gtrsim 0$$

any nett diffusion conserving μ, J must be inwards. Furthermore because a particle's energy rises as it moves to lower L with μ, J constant, particles in resonance must gain energy overall. It follows that the waves driving such diffusion are damped by the resonance $N=0$.

We can draw one very important conclusion. The expected ring current distribution normally is such that for a wave to be generated by resonant particles it should have the $N \neq 0$ resonances dominant. At the inner edge of the ring current this is particularly true because whether or not an inverted energy distribution may occur, necessarily $(\partial f/\partial L)_{\mu,J} > 0$ and nett spatial diffusion thus must be inward and upward in energy if $N=0$.

6. Symmetry

The previous section emphasised the particular nature of the hot magnetospheric plasma population. This section points out the significance of the rough North-South symmetry of the Earth's field. The lowest frequency waves have a standing structure along the field and wave fields will be either symmetric or antisymmetric about the equator if the background field is fairly symmetric. The standing structure has been confirmed for geomagnetic pulsations in several ways (see e.g. LANZEROTTI and SOUTHWOOD, 1979). Symmetry or antisymmetry in wave electric field and magnetic field means $\dot{W}$, the acceler-

ation exhibits symmetry.   If $\dot{W}$ is symmetric about the equator, terms with N odd are
missing from series (8) and so only $N=0$, ± 2 etc. resonances occur.   If $\dot{W}$ is antisymmetric,
only odd terms are present in (8) and only the $N=\pm 1$, ± 3 etc. resonances occur.   Let us
now consider a simple example, a transverse Alfvén wave.   In the simplest circumstance
where plasma pressure and inhomogeneity are unimportant the Alfvén wave has $b_{\parallel}$ and
$E_{\parallel}$ zero so

$$\dot{W}=q\boldsymbol{E}\cdot\boldsymbol{v}_{\mathrm{d}}$$

and the symmetry of $\boldsymbol{E}$ is reflected in $\dot{W}$.   The fundamental standing Alfvén wave illustrated
in Fig. 2(a) has $\dot{W}$ symmetric, the next harmonic, shown in Fig. 2(b), has $\dot{W}$ antisymmetric.

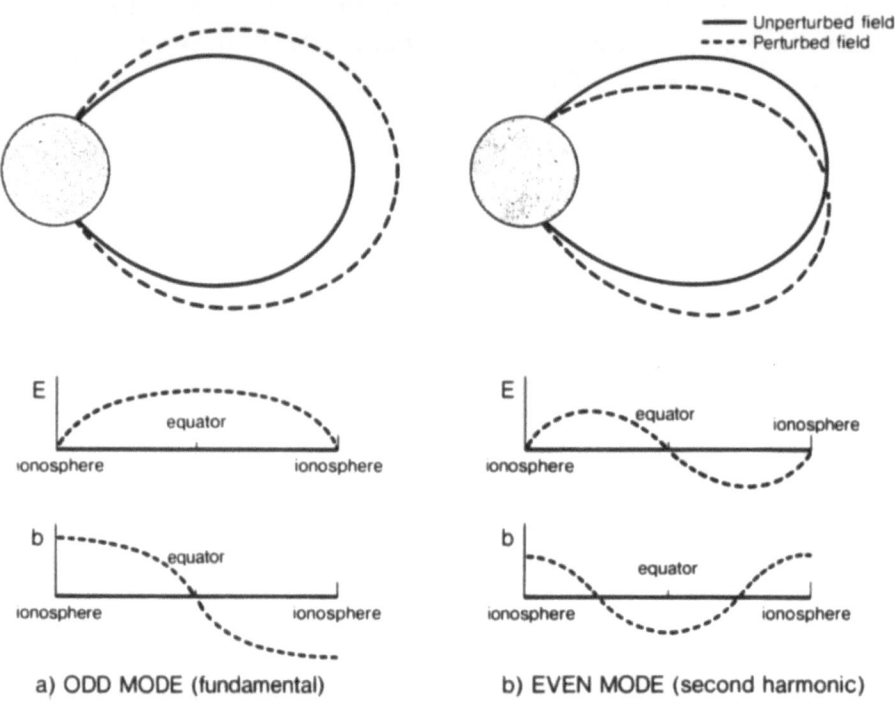

STANDING OSCILLATIONS IN A DIPOLE FIELD

Fig. 2. (a) Schematic of field line displacement (top), electric field amplitude along the
background field (middle), and magnetic amplitude (bottom) in an odd mode fundamental standing Alfvén wave. (b) As for (a) for the lowest frequency even mode
(second harmonic).

Because the parallel wavelength is long in both cases the signal varies slowly over most
particles bounce orbits and the most significant resonances are those with N smallest
so wave (a) has the $N=0$ resonance dominant and wave (b) the $N=\pm 1$.   In the $N=0$
resonance particles conserve μ, J and resonance damps the wave.   Wave (b)'s configuration is more appropriate to extract energy from the particle population.

7. Tests for Resonant Wave Generation

The results of the previous two sections suggested the fundamental Alfvén mode is most likely to be damped by resonant interactions as the $N=0$ resonance should be dominant.   One test for resonant generation would be thus to look for instances of excitation of the antisymmetric second harmonic wave illustrated in Fig. 2(b).   Its lack of symmetry about the equator would make it hard to excite by other mechanisms such as Kelvin-Helmholtz instability on the boundary (SOUTHWOOD, 1968, 1974) at least in isolation.   This argument led SOUTHWOOD et al. (1969) to suggest the transverse magnetic oscillations reported at synchrous orbit on the ATS 1 spacecraft by CUMMINGS et al. (1969) were due to bounce resonance excitation.   ATS 1 was very close to the equator and symmetry considerations suggest only the second harmonic would produce a significant transverse magnetic signal there (cf. Figs. 2(a) and (b)).   A less ambiguous determination of mode type in space requires information on the electric field or equivalently, plasma velocity or displacement.   In Fig. 2(a) above the equator the transverse magnetic perturbation is in antiphase with the field line displacement while below the equator they are in phase.   In case Fig. 2(b) precisely the reverse holds.

Electric field measurements made in space at pulsation frequencies are somewhat sparse.   Measurements of particle flux oscillations have commonly been made and under appropriate circumustances such measurements allow one to deduce the electric field (KIVELSON, 1976; KOKUBUN et al., 1977; CUMMINGS et al., 1978; HUGHES et al., 1979).   Particles of all energies respond to the wave fields.   At low enough energy the dominant response in a directional particle detector is due to the $E \times B$ drift alternately sweeping plasma into and out of the detector and the variation of flux seen is thus a measure of the electric field at right angles to the look direction of the detector.   KIVELSON (1976) recognised that many fluctuations in particle current to the OGO 5 light ion mass spectrometer could be attributed to this hydromagnetic wave effect rather than to density enhancements swept in by the spacecraft ram velocity.   Absence of information on the cold ion temperature means in this instance there is an ambiguity in the precise electric field amplitude.   This is removed if low energy particles are measured over a range of energies as in CUMMINGS et al.'s (1978) and HUGHES et al.'s (1979) work.   As HUGHES et al. (1979) show the wave $E \times B$ can dominate the behaviour of the ion flux in a detector pointing in a particular direction up to keV energies.

It is appropriate to note as an aside at this point that the acceleration which causes the flux change proportional to the $E \times B$ in a directional detector is ignored in the adiabatic acceleration expression (see e.g. (4)).   The effect is dropped in the theory because as far as each particle is concerned it is gyrophase dependent and averages to zero in a gyroperiod.   A detector with a fixed look direction samples a fixed gyrophase and thus detects the acceleration and deceleration in phase with E.

An effect with no dynamic importance thus is of great observational and diagnostic importance.   In particular on an appropriately instrumented spacecraft it can and has lead to the ability to measure E.   When b and E are measured it is usually straightforward to deduce the symmetry of the wave.   For resonant generation we expect the configuration of Fig. 2(b) to be most likely.

8. Pc 4 Signals in the Late Afternoon

There is a further type of measurement that we did not mention in the last section but which can help discriminate resonant generation. It is to measure the wave wavelength perpendicular to the field. Resonantly generated waves should have small east-west wavelengths as we argued earlier. HUGHES *et al.* (1978a) report measurements from three spacecraft in synchronous orbit. One of their most important discoveries was a null result. In the late afternoon sector during the week of their study there was usually Pc 4 pulsation activity present at all three spacecraft simultaneously but signals were incoherent between the spacecraft. The incoherence meant no wavelength determination could be made. The coherence length should exceed or be of order the wavelength and so the wavelength must be shorter than the smaller spacecraft separation. These separations varied as one spacecraft was being moved. Using HUGHES *et al.*'s (1978a) minimum figures it seems $m > 100$. Now the energy inversion in the proton population is generally clearest in precisely this same sector of local time at synchronous orbit (Whipple, 1979, private communication) and it is interesting to ask if there could be a connection. HUGHES *et al.* (1978b) reported on one particular afternoon event on February 13, 1975. For this event they had both magnetometer data (shown in Fig. 8 of HUGHES' (1980) paper in this collection) and particle data measured by the UCSD plasma instrument. Figure 3

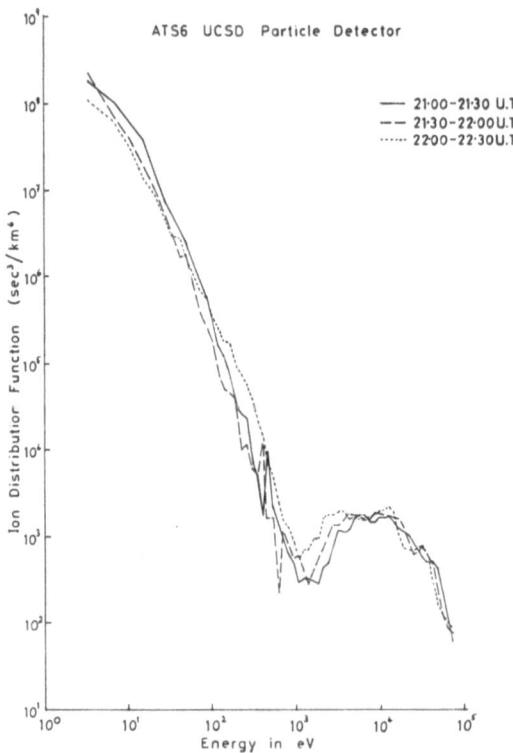

Fig. 3. Mean ion distribution function measured by UCSD ATS 6 particle detector in three half hour intervals on 16th February, 1975 during the presence of Pc 4 pulsation activity.

shows the ion (assumed to be protons) distributions for successive half-hour periods. A more detailed study of how the low energy proton count oscillated in the wave (using the phase method described by HUGHES *et al.* (1979)) showed that the magnetic oscillation perpendicular to *B* was in phase with the field displacement. ATS 6 is above the effective field line equator and so, as our earlier argument shows, the field line is oscillating in a second harmonic configuration.

The evidence thus points to a class of Pc 4 signal seen at synchronous orbit in late afternoon being resonantly generated. If *m* numbers are as high as suggested these signals may not even be observable on the ground for the amplitude of signals with rapid horizontal variation (scale at ground <100 km) are strongly attenuated between the ionosphere and the ground (HUGHES and SOUTHWOOD, 1976). As the spatial gradient in particles is outward it seems likely $\mathrm{d}W/\mathrm{d}L$ must be positive for particles to diffuse to lower energy (cf. Fig. 1). For $\mathrm{d}W/\mathrm{d}L$ to be positive for protons waves must be moving eastwards. On the other hand the protons themselves should be ∇B drifting westwards. It is unwise to attribute the "filling" of the distribution seen on the spacecraft (see Fig. 3) to a non-linear wave effect. As the spacecraft rotates towards local dusk it is moving closer to the source of protons. It seems reasonable to think of the spacecraft moving through a steady state system of interacting waves and particles. A likely scenario for wave generation is shown in Fig. 4. Convection feeds ions from the nightside into the afternoon sector. The ions drive Alfvén waves which are unstable on the dayside but

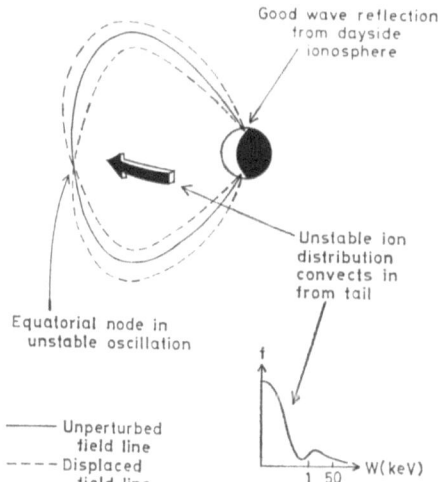

Fig. 4. Scenario for the maintenance of steady short perpendicular wavelength Pc 4 activity in the afternoon sector.

not on nightside flux tubes. Behind the dawn-dusk meridian field lines are stable because the ionospheric height integrated conductivity Σ_p is comparable to the Alfvén wave "conductivity" Σ_A ($\simeq 1/\mu_0 A$) and waves are thus badly reflected (HUGHES and SOUTHWOOD, 1976). On the dayside the ionosphere is a far better conductor and thus a far more effective reflector of h.m. wave energy and accordingly less energy is required from particles to drive the waves.

9.   Concluding Remarks

The Pc 4 events seen at synchronous orbit in the late afternoon sector seem very likely to be excited by resonant particles.   Both circumstantial evidence (occurrence in late afternoon) and structural evidence (the even mode symmetry about the equator) fit with theory.   However the story is not complete.

We have not attempted to review which wave mode is most readily generated. SOUTH-WOOD (1976) argued that in magnetospheric conditions a standing Alfvén wave would be most likely because of its low frequency.   In a hot inhomogeneous plasma the signal could be substantially modified.   In particular a compressional magnetic guided wave appears possible under appropriate conditions (SOUTHWOOD, 1977).   SOUTHWOOD (1976) did not consider waves with $E_{\parallel} \neq 0$; he assumed there was a background cold plasma of ionospheric origin adequate to suppress such a component.   The most important mode with $E_{\parallel} \neq 0$ seems to be the kinetic Alfvén wave emphasised by Hasegawa (e.g. HASEGAWA, 1979), a short perpendicular wavelength limit of the trans erse Alfvén wave.

Much of our discussion used the transverse wave as illustration although few of our arguments are tied to it.   HUGHES' (1980) Fig. 8 shows that the late afternoon Pc 4 signals have a substantial compressional component.   The wave frequency is appropriate for a standing Alfvén wave and much lower than would be expected for a localised fast, mode wave (as $\omega \propto k$ for the fast mode).   The magnetic compression in a slow mode is likely to be balanced by a particle pressure change that is in antiphase.   The fact that the ring current pressure is comparable to the background magnetic pressure means that this is entirely possible.   SOUTHWOOD (1977) gives a theory of a localised guided signal which could be appropriate.   A test is to examine the pressure oscillation in such signals. In one case where this has been done (HUGHES et al., 1979) pressures do not appear to balance; more theory and data analysis should be done.

This work was supported in part by NSF under contract ATM 79–23586 at the University of California Los Angeles.

REFERENCES

ATKINSON, G. and T. WATANABE, Surface waves on the magnetospheric boundary as a possible origin of long period geomagnetic micropulsations, Earth Planet. Sci. Lett., 1, 89 91, 1966.

CHEN, A. J., Penetration of low energy protons deep into the magnetosphere, J. Geophys. Res., 75, 2458–2467, 1970.

CHEN, L. and A. HASEGAWA, A theory of long-period magnetic pulsations, 1, Steady state excitation of field line resonance, J. Geophys. Res., 79, 1024–1032, 1974.

CLEMMOW, P. C. and J. P. DOUGHERTY, Electrodynamics of Particles and Plasmas, Addison-Wesley, London, 1969.

COWLEY, S. W. H., Energy transport and diffusion, in Physics of Solar Planetary Environments, Vol. 2, edited by D. J. Williams, 582 pp., Am. Geophys. Union, Washington D.C., 1976.

COWLEY, S. W. H. and M. A. M. ASHOUR-ABDALLA, Adiabatic plasma convection in a dipole field: Proton forbidden-zone effects for a simple electric field model, Planet. Space Sci., 24, 821–833, 1976.

CUMMINGS, W. D., R. J. O'SULLIVAN, and P. J. COLEMAN, Jr., Standing Alfvén waves in the magnetosphere, J. Geophys. Res., 74, 778–793, 1969.

CUMMINGS, W. D., S. E. DeFOREST, and R. L. McPHERRON, Measurements of the Poynting vector of standing hydromagnetic waves at geosynchronous orbit, J. Geophys. Res., 83, 697–706, 1978.

DUNGEY, J. W., Electrodynamics of the outer atmosphere, Report 69, 33 pp., Ionos. Res. Lab., Pa. State Univ., 1954.

DUNGEY, J. W., Survey of acceleration and diffusion, in *Radiation Trapped in the Earth's Magnetic Field*, edited by B. M. McCormac, 389 pp., D. Reidel, Dordrecht, Netherlands, 1966.

DUNGEY, J. W. and D. J. SOUTHWOOD, Ultra low frequency waves in the magnetosphere, *Space Sci. Rev.*, **10**, 672–688, 1970.

GREEN, C. A., The longitudinal phase variation of Pc 3–4 micropulsations, *Planet. Space Sci.*, **24**, 79–86, 1976.

HAMLIN, D. A., R. KARPLUS, R. C. VICK, and K. M. WATSON, Mirror and azimuthal drift frequencies for geomagnetically trapped particles, *J. Geophys. Res.*, **66**, 1–4, 1961.

HASEGAWA, A., Particle dynamics in low frequency waves in an inhomogeneous plasma, *Phys. Fluids*, **22**, 1988–1993, 1979.

HASEGAWA, A. and K. MIMA, Anomalous transport produced by kinetic Alfvén wave turbulence, *J. Geophys. Res.*, **83**, 1117–1124, 1978 (correction: *J. Geophys. Res.*, **83**, 5778, 1978).

HERRON, T. J., Phase characteristics of geomagnetic pulsations, *J. Geophys. Res.*, **71**, 871–889, 1966.

HUGHES, W. J., Multisatellite observations of geomagnetic pulsations, *J. Geomag. Geoelectr.*, **32** Suppl. II, 1980 (this issue).

HUGHES, W. J. and D. J. SOUTHWOOD, The screening of micropulsation signals by the atmosphere and ionosphere, *J. Geophys. Res.*, **81**, 3234–3240, 1976.

HUGHES, W. J., R. L. MCPHERRON, and J. N. BARFIELD, Geomagnetic pulsations observed simultaneously on three geostationary spacecraft, *J. Geophys. Res.*, **83**, 1109–1116, 1978a.

HUGHES, W. J., D. J. SOUTHWOOD, B. H. MAUK, R. L. MCPHERRON, and J. N. BARFIELD, Alfvén waves generated by an inverted plasma energy distribution, *Nature*, **275**, 43–45, 1978b.

HUGHES, W. J., R. L. MCPHERRON, J. N. BARFIELD, and B. H. MAUK, A compressional Pc 4 pulsation observed by three satellites in geostationary orbit near local midnight, *Planet. Space Sci.*, **27**, 821–840, 1979.

KIVELSON, M. G., Instability phenomena in detached plasma regions, *J. Atmos. Terr. Phys.*, **38**, 1115–1126, 1976.

KIVELSON, M. G. and D. J. SOUTHWOOD, Approximations for the study of drift boundaries in the magnetosphere, *J. Geophys. Res.*, **80**, 3528–3534, 1975.

KOKUBUN, S., M. G. KIVELSON, R. L. MCPHERRON, C. T. RUSSELL, and H. I. WEST, Jr., OGO 5 observations of Pc 5 waves: Particle flux modulations, *J. Geophys. Res.*, 2774–2786, 1977.

KOZHEVNIKOV, A. A., A. B. MIKHAILOVSKY, and O. A. POKHOTELOV, The role of protons of the radiation belts in the generation of Pc 3–5, *Planet. Space Sci.*, **24**, 465–474, 1976.

LANZEROTTI, L. J. and D. J. SOUTHWOOD, Hydromagnetic waves, in *Solar System Plasma Physics*, Vol. 3, edited by C. F. Kennel, L. J. Lanzerotti, and E. N. Parker, 109 pp., North Holland, Amsterdam, 1979.

LANZEROTTI, L. J., H. FUKUNISHI, and L. CHEN, U.L.F. pulsation evidence at the plasmapause 3. Interpretation of polarisation and spectral amplitude studies of Pc 3 and Pc 4 pulsations near $L=4$, *J. Geophys. Res.*, **79**, 4648–4653, 1974.

MAUK, B. H. and C. E. MCILWAIN, Correlation of K_p with the substorm injection boundary, *J. Geophys. Res.*, **79**, 3193, 1974.

MIER-JEDRZEJOWICZ, W. A. C. and D. J. SOUTHWOOD, The east-west structure of mid-latitude geomagnetic pulsations in the 8–25 mHz bond, *Planet. Space Sci.*, **27**, 617–630, 1979.

NORTHROP, T. G., *The Adiabatic Motion of Charged Particles Interscience*, 109 pp. New York, 1963.

NORTHROP, T. G. and E. TELLER, Stability of adiabatic motion of charged particles in the Earth's field, *Phys. Rev.*, **117**, 215–225, 1960.

OLSON, J. V. and G. ROSTOKER, Longitudinal phase variations of Pc 4–5 micropulsations, *J. Geophys. Res.*, **83**, 2481–2488, 1978.

SMITH, P. H. and R. A. HOFFMAN, Direct observations in the dusk hours of the characteristics of the storm time ring current particles during the beginning of magnetic storms, *J. Geophys. Res.*, **79**, 966–971, 1974.

SOUTHWOOD, D. J., The hydromagnetic stability of the magnetospheric boundary, *Planet. Space Sci.*, **16**,

587–606, 1968.

SOUTHWOOD, D. J., The behaviour of ULF waves and particles in the magnetosphere, *Planet. Space Sci.*, **21**, 53–65, 1973.

SOUTHWOOD, D. J., Some features of field line resonances in the magnetosphere, *Planet. Space Sci.*, **22**, 483–491, 1974.

SOUTHWOOD, D. J., A general approach to low frequency instability in the ring current plasma, *J. Geophys. Res.*, **81**, 3340–3348, 1976.

SOUTHWOOD, D. J., Localised compressional hydromagnetic waves in the magnetospheric ring current, *Planet. Space Sci.*, **25**, 549–554, 1977.

SOUTHWOOD, D. J. and W. F. STUART, Pulsations at the substorm onset, in *Dynamics of the Magnetosphere*, edited by S.-I. Akasofu, Riedel, Dordrecht, Holland, 1979.

SOUTHWOOD, D. J., J. W. DUNGEY, and R. G. ETHERINGTON, Bounce resonant interaction between pulsations and trapped particles, *Planet. Space Sci.*, **17**, 349–361, 1969.

SOUTHWOOD, D. J., R. J. ETHERINGTON, and J. W. DUNGEY, Neglected plasma instability involving bounce resonance, *Nature*, **219**, 56–57, 1968.

Solar Wind Control of Daytime, Midperiod Geomagnetic Pulsations

E. W. Greenstadt,* R. L. McPherron,** and K. Takahashi**

*Space Sciences Department, TRW Defense and Space Systems Group,
Redondo Beach, California, U.S.A.
**Institute of Geophysics and Planetary Physics, University of California,
Los Angeles, California, U.S.A.

(Received June 28, 1980)

Numerous studies have established that various properties of geomagnetic pulsations are linked to various properties of the solar wind. The linkage in most cases is rather loose, suggesting that combinations of factors must be involved in generation and control of pulsation activity. We review briefly the most significant observational results and we describe and discuss critically the two most prominent models for external generation of magnetospheric waves. We present arguments favoring joint application of the models, wherein perturbations in the magnetosheath resulting from favorable interplanetary field orientation are delivered to the magnetopause, transferred directly into the subsolar magnetosphere, and amplified into surface waves on the flank of the magnetosphere by Kelvin-Helmholtz instability at high solar wind speed. This combination of circumstances can account for experimental correlations of pulsation occurrence with interplanetary field orientation, periods with interplanetary field strength, and amplitudes with solar wind velocity.

1. Introduction

More than a decade of observational studies have accumulated a growing library of evidence that daytime midperiod geomagnetic pulsations appear and disappear, or, more accurately, grow, persist, and shrink in accordance with the state of the solar wind. The pulsations to which we refer are those in the range of periods T traditionally covered by Pc 3 ($T=10$ to 45 s), Pc 4 ($T=45$ to 150 s), and Pc 5 ($T=150$ to 300 s).

Evidence has thus been mounting that these geomagnetic pulsations are controlled, if not actually generated, by the solar wind. Many parameters of the solar wind have been correlated with various attributes of pulsations. Although few of the published results have been completely definitive, the cumulative effect of many independent investigations with diverse data sets provides persuasive evidence that the effects are real.

The earliest work linking pulsations to the solar wind was done in the U.S.S.R. and has been reviewed by Gul'elmi (1974). Brief reviews of more recent studies have been given by Saito et al. (1979) and Greenstadt (1979). Rather than present a new, detailed review of all experimental material published to date, we have interpreted our charter mainly as an opportunity to discuss the proposed explanations for solar wind generation of pulsations. We think this is important at this time because we see a need for new studies that should be undertaken at the earliest opportunity.

Table 1. Locations of observatories used in studies of pulsations and solar wind, and quantities that were correlated.

Investigators	Magnetic latitude: 0 / L-value: 1	45 / 2	55 / 3	60 / 4	63 / 5	66 / 6	68 / 7	69 / 8	Occurrence	Amplitude	Frequency
									Pulsation characteristics — Correlated quantities		
BOL'SHAKOVA and TROITSKAYA (1968)		G								IMF_{long}	V_{sw}
TROITSKAYA et al. (1969)		G	G						B, V_{sw}		
GRINGAUZ et al. (1971)		G	G								N
TROITSKAYA et al. (1971)									IMF_{long}	IMF_{long}	B, IMF_{long}
GUL'ELMI et al. (1973)		G	G							V_{sw}	B
GUL'ELMI and BOL'SHAKOVA (1973)		G	G								B, K_p, A_e
VINOGRADOV and PARKHOMOV (1974)		G								V_{sw}	
GREENSTADT and OLSON (1976)			G							IMF_{rad} (θ_{XB})	
KOVNER et al. (1976)		G	G	G							B
NOURRY (1976)							ATS-1		IMF_{rad}	IMF_{rad}	Fu.w.
RUSSELL and FLEMING (1976)			G	G							B
WEBB and ORR (1976)			G	G						IMF_{rad}	
ARTHUR and McPHERRON (1977)						ATS-6				IMF_{rad}	
GREENSTADT and OLSON (1977)			G	G						IMF_{rad}	
SINGER et al. (1977)			G	G						V_{sw}	
WEBB et al. (1977)			G							IMF_{rad}	
PLYASOVA-BAKOUNINA et al. (1978)				G						IMF_{rad}	B, Fu.w.
VERÓ and HOLLÓ (1978)		G								IMF_{rad}, K_p	B
GREENSTADT et al. (1979a)				G						IMF_{rad}, V_{sw}	
SAITO et al. (1979)	G									IMF_{rad}, V_{sw}, K_p	
GREENSTADT and OLSON (1979)			G							IMF_{rad}	
GREENSTADT et al. (1979b)			G	G						V_{sw}	
WOLFE et al. (1980)			G							V_{sw}, IMF_{rad}	
WOLFE (1980)			G							V_{sw}, IMF_{rad}	

G, Ground stations; ATS, Geosynchronous satellites; Fu.w., Frequency of upstream waves.

In the following remarks we summarize the observational studies tying the solar wind to pulsations seen at the ground, and describe the most convincing and prominent evidence relating the solar wind to pulsations, including some new results with multiple parameter correlations and waves at geostationary altitude. We then examine the two principal models of solar wind control of pulsations, evaluate them critically, and enumerate a few studies aimed at exploring the models further. Finally, we conclude by advocating development of a global pulsation index.

2. Observations

2.1 Perspective

Table 1 gives a broad overview of the locations on the ground and in the magnetosphere where pulsations have been correlated with solar wind parameters, plus the parameters which have been correlated and the principal reporters of those correlations. We see that the overwhelming majority of studies have recorded pulsations between about $L \approx 1.8$ and $L \approx 3.5$. The earlier studies concentrated on observations between 1.8 and 2.8 L, the later ones on observations around 3.5 L plus the satellite distance of 6.6L. It can hardly be said that the magnetic latitudes have been widely, or systematically, explored, but the extremes of position are spread sufficiently to encourage confidence in the generality of the implied connection between solar wind conditions and waves throughout the dayside magnetosphere. Table 2 summarizes the relations discovered so far for the Pc 3 band, which has been the most consistently tested by all observers.

Table 2. Observational relations between solar wind parameters and Pc 3 pulsations.

* Occurrence and/or amplitude

    1. Negative correlation with THETA$_{XB}$ (IMF$_{rad}$).
    2. No Pc 3 when IMF $\perp$ sun-earth line.
    3. Positive correlation with V_{SW}.
    4. Positive correlation with K_p index.

* Period
    1. Negative correlation with B: $T \propto B^{-1}$.
    2. Positive correlation with N.
    3. Same as the upstream waves.

* Others
    1. More occurrence when $B_x > N$.
    2. Little is known about polarization ellipticity latitudinal dependence.

There are two inconstant aspects of this subject not apparent in Table 1 that bear emphasis from a purely experimental point of view. First, the earlier studies were selective and one-sided: That is, the observers selected events on the ground clearly identifiable at nearly monochromatic Pc 3 or Pc 4 wavetrains and then tabulated the corresponding solar wind variables. In contrast, the later studies have increasingly concentrated on more quantitative measures of pulsation "activity," whether identifiable as classic Pc wavetrains or not, and have compared measures of pulsation energy with interplanetary parameters, without preselection on either side. Second, the general quality and "processibility" of available data, both from ground stations and satellites, have improved greatly through the 12 years spanned by the table.

These considerations bring into question the uniformity that can be attached to the phenomena tested by the efforts summarized in Tables 1 and 2. Perhaps the occurrence of definable wavetrains and the amplitude of band energy are wholly separate phenomena, and the early and late correlations have quite different meanings. We are inclined to the view that, while two or more distinguishable sources may contribute to generation of Pc signals, the control of such signals by the solar wind is a consistent theme of all the results since the first investigation, and that, allowing for some reinterpretation through hindsight, wavetrains identified in some studies and broadband signal level measured in others were manifestations of a common phenomenon.

The following paragraphs and figures illustrate some of the relationships indicated in Tables 1 and 2. They also portray the quality of current observational results and convey, by virtue of the comparitive feebleness of individual correlations, the reasons we cited persistence of relations from study to study as a measure of credibility.

2.2   Illustrations

2.2.1   IMF magnitude and period

The most persuasive correlation yet found between a property of the solar wind and geomagnetic pulsations at the ground has been that of the IMF magnitude B and pul-

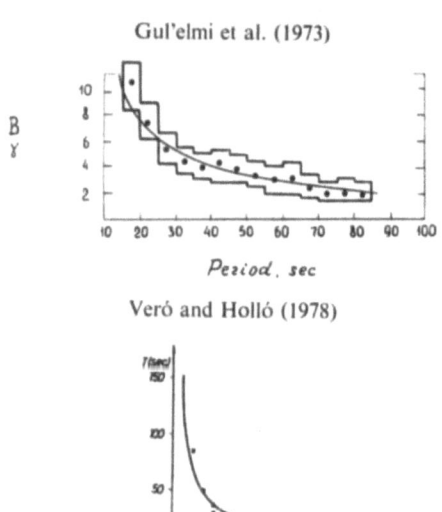

Fig. 1. Two versions of the observational results relating B in the solar wind to the period T of daytime pulsations. The curves are drawn from $T = 160/B$.

sation period T. Figure 1 shows two versions of the correlation, one developed by Gul'elmi *et al.* (1973), the other by Veró and Holló (1978). The curves through the data points in both cases are represented by T (sec) $= 160/B$ (γ). The apparent close agreement between the data points and the relationship represented by the curves has been qualified by Veró (1979), who cautioned that the correlation is a "very rough one", with

T really representing the period of peak activity over a band of excited frequencies. Other qualifications have also appeared, generally in the form of modifications of the above expression, by changing the proportionality constants (GUL'ELMI, 1974) and by adding a linear term in K_p to account for a general increase of pulsation activity with geomagnetic disturbance. One such composite representation served as the basis for the "Borok index", an hourly number designed to provide an indirect measure of the IMF from pulsation data. The formula used was $B=0.7+16K_p+150/T$.

The reliability of the Borok index was tested by RUSSELL and FLEMING (1976), who compared the index with actual measurements of the IMF by spacecraft. Their results gave a spread of measured field strengths corresponding to each index value B, as displayed in Fig. 2. Clearly, the peak of every distribution did not correspond to the associated index value, and Russell and Fleming observed that the index could be improved by a suitable displacement. They devised such a "recalibration" and concluded that when so corrected, their revised Borok index was an appreciably better predictor of the IMF than random numbers, but still fell far short of providing a one-for-one substitute for direct measurement.

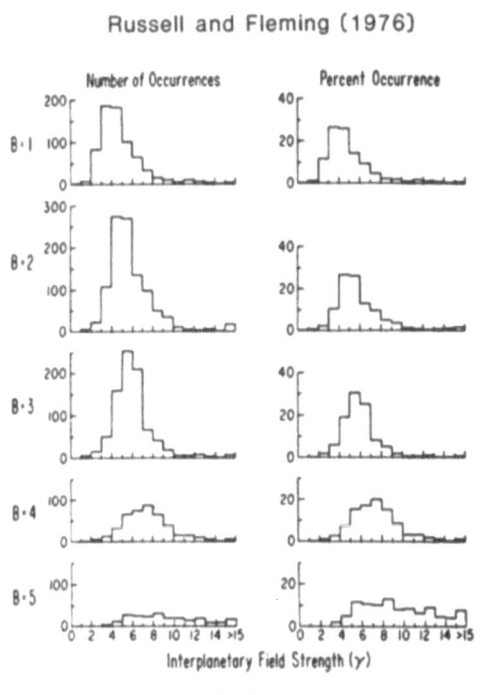

Fig. 2

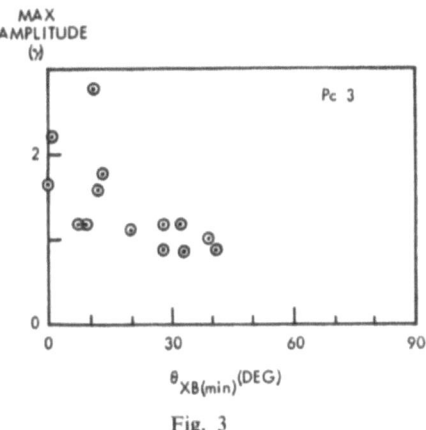

Fig. 3

Fig. 2. Histograms of the occurrence rate of interplanetary field strength for $1-\gamma$ bins.

Fig. 3. Scatter plot of Pc 3 hourly data selected because they violate criteria for Pc 3 appearance based on values of either K_p or the IMF.

Subsequently, GREENSTADT and OLSON (1977) used an expression for the Borok index, revised according to the Russell and Fleming correction, to select those hours for which the IMF and K_p values indicated pulsation wavetrains should have been excited *outside* the Pc 3 range. They found, however, that the hourly maximal amplitudes *in* the Pc 3 band, Fig. 3, exhibited much the same trend with IMF cone angle θ_{XB} (see sections following) that unselected data did, suggesting that Pc 3 *signals* were present after all.

It appears at present that, given Veró's "very rough" correlation, Russell and Fleming's broad, overlapping, and displaced correspondences, and Greenstadt and Olson's "unauthorized" correlation, there is not yet an established, quantitatively reliable relationship between IMF magnitude and pulsation period, except in a statistical sense.

2.2.2   Speed and cone angle at the ground

Figure 4 depicts the most prominent solar wind variables recently correlated with ground-level pulsation activity. The scatter diagrams also illustrate the quality of the known correlations of solar wind velocity and cone angle with pulsations in the Pc 3 and Pc 4 ranges, while the vertical scales indicate two of the representations of pulsation activity that have been used: The data at far left are based on occurrence, while those of the center and right are based on amplitude, in this case the hourly maximal amplitude of filtered signals.

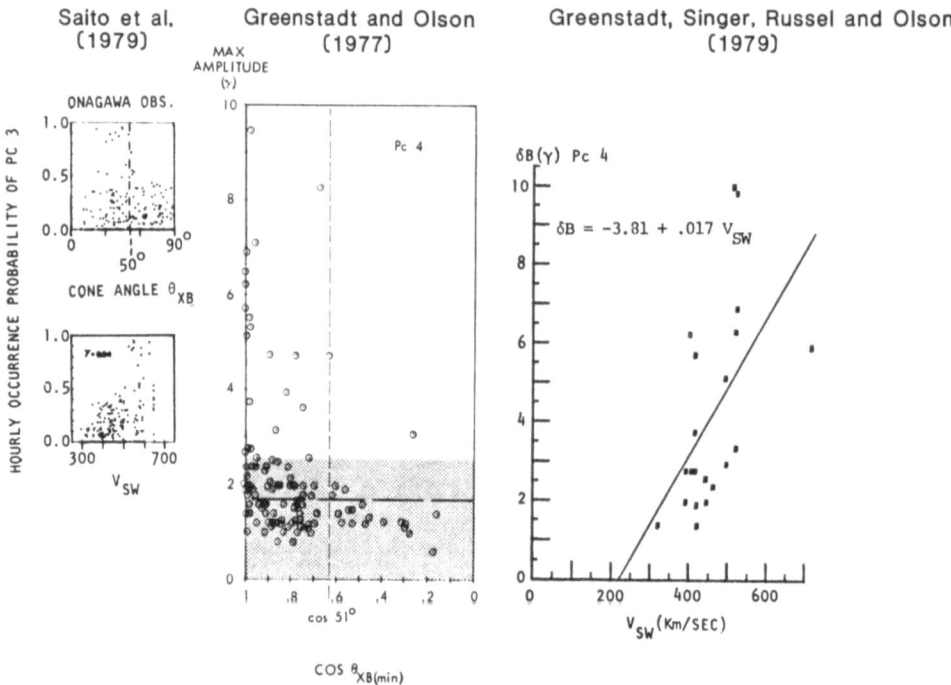

Fig. 4.   Solar wind speed and IMF cone angle dependence of Pc 3, 4 occurrence.

The two sets of data show, first, that pulsation activity increases with lower IMF cone angle and higher solar wind speed, whether measured by probability of occurrence or by actual signal amplitude; second, that both Pc 3 and Pc 4 period-ranges are affected similarly by these solar wind parameters; third, that the correlations are trends discernible mostly in the upper envelope of the data amid appreciable scatter, indicating severe interference by either methodological weakness, competing variables, or both; fourth, that cone angles less than about 50° are particularly favorable to the occurrence and amplitude of the largest Pc signals.

Note that the amplitude correlations have also been developed for Pc 3 (GREENSTADT and OLSON, 1977) and, with respect to V_{SW}, for Pc 5 (GREENSTADT et al., 1979a).

One attempt to separate the influence of solar wind variables on ground-level pulsation activity has been made by WOLFE et al. (1979, 1980). WOLFE (1980) has also lately replaced visually defined correlations with more rigorous multivariate analysis. Estimates of power spectral density were used to obtain quantitative measures of hourly pulsation activity in the period ranges 31–60 sec, 60–120 sec, and 120–240 sec. Hourly average values of leading solar wind parameters were also examined to determine which might be sufficiently uncorrelated to serve as truly independent measures of solar wind state; solar wind speed and IMF cone angle were the best candidates.

The results of Wolfe et al. paralleled those cited in the preceding paragraphs for Pc 3 and Pc 4. These authors found, however, that the correlation of pulsation activity with cone angle became weak or nonexistent for the longest periods (120–240 sec), although the correlation with solar wind speed remained strong.

The correlation plots of WOLFE (1980) are much the same as those already shown, containing much scatter, but the multivariate analysis table, Fig. 5, is an important improvement in method of displaying the results of such pulsation studies. The figure shows that the addition of parameters to the analysis one-by-one, reading the columns right to left, raises the multiple correlation coefficient R, first row, while reducing the residual error, second row. However, the largest changes occurred with the addition of θ_{XB} to V_{SW}, second column from right, with relatively little improvement introduced by

WOLFE (1979,1980)

MULTIPLE R	0.69	0.69	0.69	0.66	0.66	0.65	0.64	0.56
RESIDUAL STANDARD ERROR	0.40	0.40	0.40	0.42	0.42	0.42	0.42	0.46
SOLAR WIND VELOCITY	▓	▓	▓	▓	▓	▓	▓	
$\theta_{X\hat{B}}$	▓	▓		▓	▓	▓	▓	
IMF GSM COMPS / B_z	–	–	–	–	–	–		
B_y	–	–	–	–	–			
B_x								
DENSITY	▒	▒	▒					
B_{AVE}	–	–						
TEMPERATURE								
INTERCEPT	▓	▓	▓	▓	▓	▓	▓	▓
INTERCEPT VALUE	-3.6	-3.7	-3.7	-3.3	-3.3	-3.3	-3.3	-3.9

T = 30.72 – 60 sec

Fig. 5. Results of multiple linear regression analyses correlating solar wind parameters with pulsation spectral power density in the designated period interval, introducing parameters one at a time.

the remaining parameters.   The shading, in four grades, represents the stability of the slope of each parameter; hence its importance, to each fit, with the darkest shading standing for the greatest importance.   The reader is referred to the description of WOLFE (1980) for further explanation; we note here the value of employing multiple linear regression and the clear dominance of V_{SW} and θ_{XB} over ground pulsations in the period range 31–60 sec, which spans the traditional demarcation (45 sec) between Pc 3 and Pc 4.

Takahashi et al. (1980)

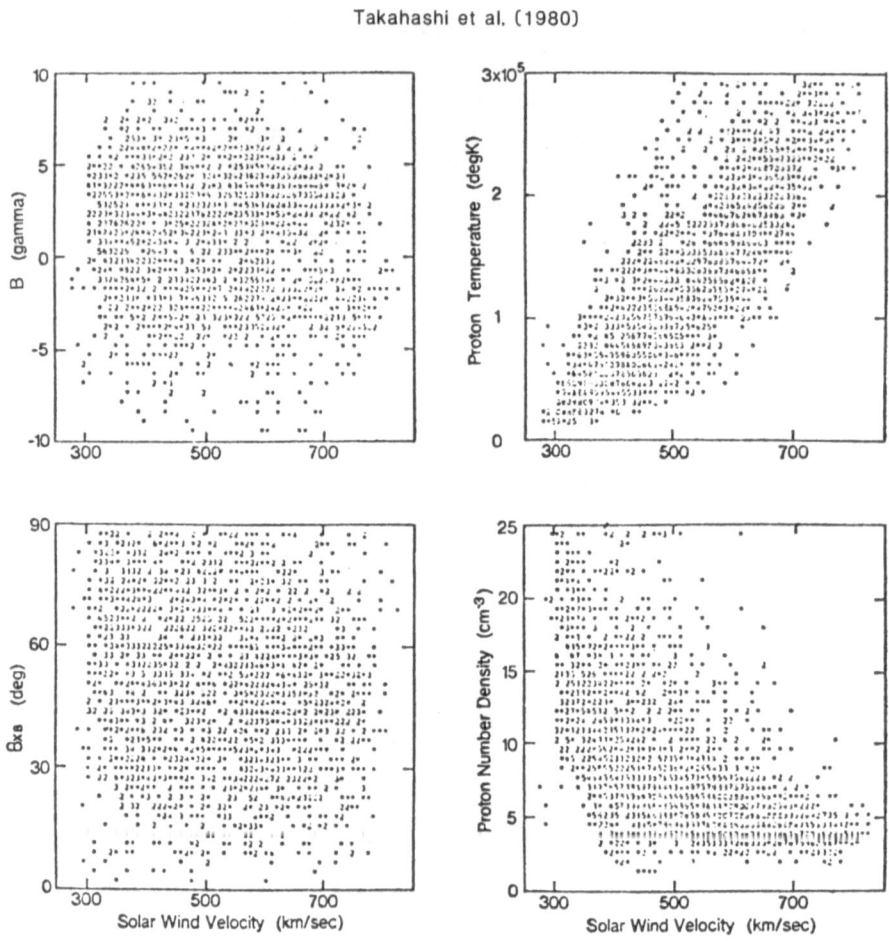

Fig. 6.   Correlation between 2-hour average solar wind parameters May 1974–May 1975.

2.2.3 Speed and cone angle at synchronous elevation

Since the phenomena of interest are magnetospheric waves, of which traditional micropulsations at the earth's surface are only one manifestation, an additional approach to tying the magnetosphere's oscillations to the solar wind is through satellite measurements.   Those made in geosynchronous orbit constitute a readily available data set, from which some new results have been compiled by TAKAHASHI et al. (1980).   The data base was some 3,000 two-hour intervals of ATS-6 magnetometer recordings between

June and September 1974.

The question inevitably arises of whether the correlations of pulsation activity with V_{SW} and θ_{XB}, already found at the ground for hourly averages, are independent, or whether V_{SW} and θ_{XB} are themselves so tightly connected over two-hour intervals that correlation of activity with one is already equivalent to correlation of activity with the other. For that matter, still other solar wind parameters might be important, as Wolfe (cited above) investigated. An inter-comparison of solar wind parameters was therefore made first.

Figure 6 shows the dependence of the X-component of magnetic field strength B_{SXE}, the angle between the sun-earth line and the IMF, θ_{XB}, plasma temperature, and plasma density, on plasma bulk velocity V_{SW}. The solar wind parameters were averaged over two-hour intervals using data compiled by J. H. King. Temperature and density are strongly correlated with V_{SW}, while θ_{XB} has little correlation with V_{SW}. So, it is reasonable to choose V_{SW} for one independent parameter, also representative of temperature and density, and θ_{XB} for another. Field magnitude B is not correlated with either V_{SW} or θ_{XB}. However, it turned out that the effect of B is smaller than that of V_{SW} or θ_{XB}. Therefore, we focus on the effect of V_{SW} and θ_{XB} in the following part, just as Wolfe *et al.* did.

Figure 7 shows that the signal power in the Pc 3 frequency range at ATS-6 increased with V_{SW}, just as the hourly maximal amplitude does on the ground.

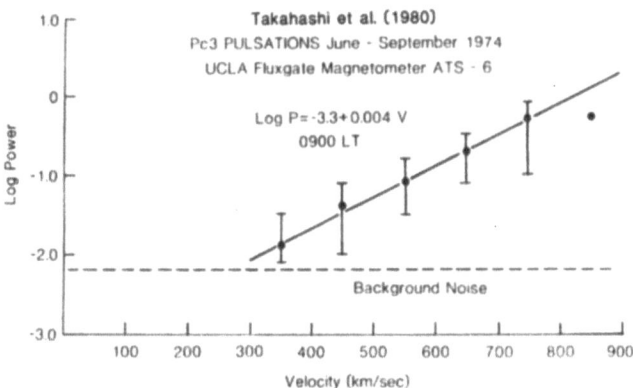

Fig. 7. Dependence of median log power on solar wind velocity.

A correlation with velocity suggests that pulsation occurrence should be connected in some way with the stream structure of the solar wind. This does seem to be the case, as shown in Fig. 8, where the solar wind velocity for two solar rotations (# 1926 and 1927), upper panel, and the corresponding Pc 3 occurrence probability, lower panel, are plotted vs. time.

Figure 9 summarizes the combined effects of both V_{SW} and θ_{XB}. The occurrence probability of Pc 3 at 6–17 LT was taken as a function of these two parameters and is shown as a contour map in the V_{SW}, θ_{XB} plane. V_{SW} and θ_{XB} have ranges of 300–800 km/sec, and 0–90°, respectively. In these ranges, V_{SW} and θ_{XB} contribute to the occurrence of Pc 3 micropulsations with equal significance. The combined effects of V_{SW} and θ_{XB}

Takahashi et al. (1980)

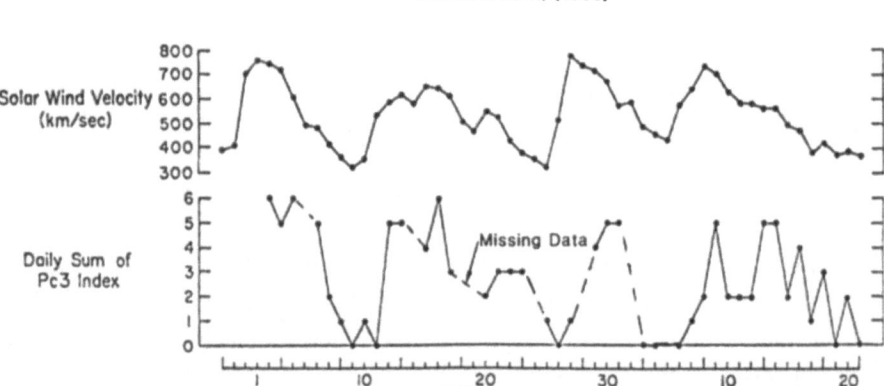

Bartels Solar Rotation 1926-1927

Fig. 8.   Recurrent pattern of solar wind parameters and occurrence of Pc 3 pulsations at synchronous orbit.

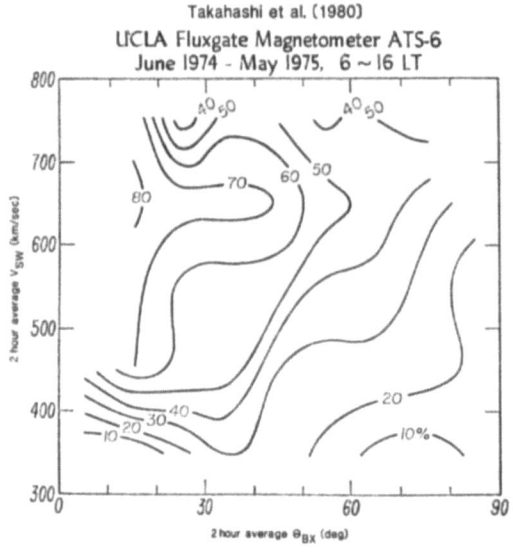

Fig. 9.   Solar wind velocity and IMF angle dependence of occurrence probability for Pc 3 magnetic pulsations.

decrease the probability to $\sim 10°$ for $V_{SW} \sim 350$ km/sec, $\theta_{XB} > 80°$, and increase it to $\sim 80\%$ for $V_{SW} \sim 650$ km/sec, $\theta_{XB} < 20°$. The peak in the contour map at $V_{SW} \sim 650$ km/sec is due to the afternoon events which have a maximum occurrence probability of $V_{SW} \sim 650$ km/sec.

In agreement with the studies by NOURRY (1976), and by ARTHUR and MCPHERRON (1977), it was also found that positive B_x was favorable for Pc occurrence.

The large central diagram in Fig. 10 summarizes the data from ATS-6 in Fig. 9 as a three-dimensional diagram. The vertical scale represents the probability of occurrence of Pc 3, based on the appearance of definable power peaks in the magnetic field spectra.

The inserts in the figure reproduce the three-dimensional summary diagrams of GREEN-STADT *et al.* (1979a), showing the general trend of the joint distribution of hourly pulsation maximum with hourly average V_{SW} and the cosine θ_{XB}.

We see that one result of the ATS study has been a remarkable verification of the joint dependence of pulsation activity on V_{SW} and θ_{XB}. The same pattern of pulsation activity rising with solar wind speed and cosine of the cone angle appears at synchronous orbit as on the ground, despite the somewhat different means of processing the data. The dependence of δB on V_{SW} is much steeper, and its occurrence much higher at low angle than high; the dependence of δB on $\cos \theta_{XB}$ is more pronounced, and the occurrence much higher at high velocity than at low. There is a hint in the ATS diagram that Pc 3 occurrence peaks or levels off at $V_{SW} \simeq 600$–700 km/sec, but a data set containing many more of the rather unusual speeds above 700 km/sec will be necessary to sustain such an inference.

We now turn to a critical examination of models developed to explain how pulsations at the ground and waves in the magnetosphere might be produced by processes external

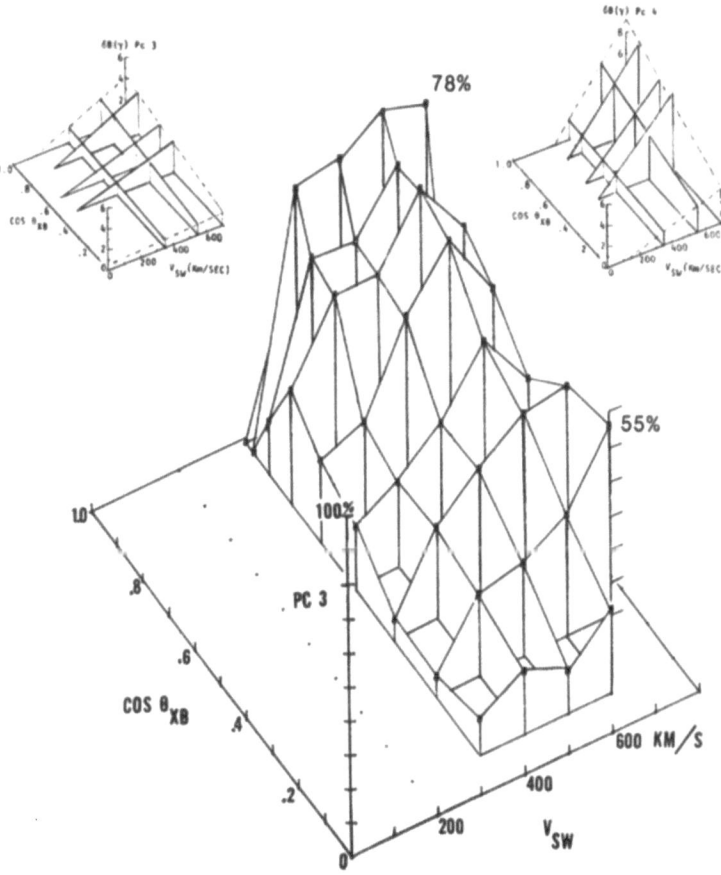

Fig. 10.   Diagrams of joint distributions of pulsations with V_{SW}, θ_{XB} for ATS-6 data (center) and on the ground (inserts).

to the magnetosphere, by which we mean the magnetopause or its plasma (solar wind) environment.

3.   Models for Solar Wind Control of ULF Pulsations

It is important first to recognize the deliberate choice of the term "control," as opposed to "generation," in the title and foregoing text of this report.   Magnetospheric ULF waves could be excited outside the magnetosphere and transferred inside, or they could be excited inside by conditions or mechanisms only controlled by the changing solar wind; for example, by pressure-controlled magnetospheric dimensions.

At present, there is no comprehensive theory of internal generation that includes external control compatible with observations.   D'Angelo (1975) postulated an endogenic mechanism that would yield a relationship between T and $1/B$, but his model did not provide for the other experimental dependencies on cone angle and solar wind velocity. We favor models of external excitation, but there has been no direct proof yet of where the wave source actually lies (Greenstadt, 1979).   We have, therefore, refrained from crediting the solar wind with more than "control" of pulsations, even though the two principal models of solar wind control are models of external generation too.

Explanations for the external origin of pulsations divide into two categories: Origin at the magnetopause and origin beyond the magnetopause.   There is presently only one model strongly advocated in each category, but others are not excluded.   Origin at the magnetopause is attributed to amplification of surface waves via a Kelvin-Helmholtz instability (Southwood, 1968; Boller and Stolov, 1973); origin beyond the magnetopause is attributed to large amplitude waves arising in the quasi-parallel bow shock, swept back into the magnetosheath and penetrating the magnetopause (Greenstadt, 1972).   Neither model necessarily eliminates the other, so their joint action is possible (Greenstadt et al., 1979a), and both depend on parameters of the solar wind.   The key parameter for the Kelvin-Helmholtz model is solar wind speed; the key parameter for wave penetration is IMF cone angle.

3.1   Descriptions of the models

3.1.1   Kelvin-Helmholtz (K-H) instability

A surface separating two fluids can be excited to produce growing waves if one fluid flows along the surface with sufficient velocity relative to the other.   If this principle is applied to the "surface" of the magnetopause, remembering that the solar wind has a stagnation zone near the subsolar point, we may obtain the situation sketched in Fig. 11, where waves are sustained along both flanks of the boundary separating magnetosheath from magnetosphere.   The instability grows when the relative velocity (effectively, the sheath velocity in this case) is above a threshold speed.   Among other things, the threshold depends inversely on the angle between the field and the velocity in the sheath flow (Boller and Stolov, 1973).   Computer simulations of the magnetosphere, in both two and three dimensions, have produced fluttering of the magnetopause that may be interpreted as caused by the K-H process (Leboeuf et al., 1979, 1980), and phenomena with the characteristics of waves have been observed at or near the magnetopause (Kaufmann et al., 1970;

BOLLER and STOLOV, 1973).   Once surface waves are established, they would presumably be transferred into and through the magnetosphere, by a process such as that suggested by SOUTHWOOD (1968, 1974) or CHEN and HASEGAWA (1974).   Waves of any given period would, in traversing the magnetosphere, undergo an enhancement of amplitude whenever they crossed a field line, or flux tube, resonant at their period.

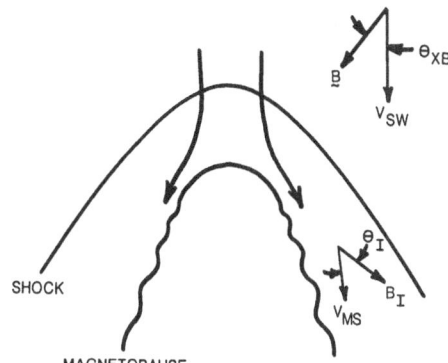

Fig. 11. Kelvin-Helmholtz stimulation of symmetrically located surface waves on the flanks of the magnetosphere.

3.1.2   Bow shock induced turbulence

Waves generated in a turbulent bow shock are swept downstream in the magnetosheath flow.   If they can propagate through the magnetopause before they are swept downstream, they become a potential source of energy for excitation of magnetospheric waves and resonant field lines.

It has been observed that the bow shock is most turbulent at locations where the magnetic field is parallel or antiparallel to the shock normal.   In the usual situation of a 45° spiral field, this condition is met at about 0900 local time.   In fact, it is found that some turbulence exists for all angles θ_{Bn} between B and $\hat{n}$ up to about 50°, but is greatest when $\theta_{Bn} \lesssim 20°$.   Thus, in the most common situation, the turbulent, "pulsation," or quasi-parallel shock structure prevails on the morning side to about the subsolar point of the bow shock (GREENSTADT and FREDRICKS, 1979).   It is known, however, that pulsations are not always present in the magnetosphere, although they are almost always present somewhere in the magnetosheath, so it is necessary to account for the variability of pulsation occurrence or signal level.

If we note that small angles θ_{XB} between the field and velocity of the solar wind move the pattern of maximal turbulence toward local noon, and that the only solar wind plasma actually making contact with the magnetopause is in the flow tube through the subsolar shock (SPREITER and ALKSNE, 1969), we expect larger waves to illuminate the magnetopause when cone angle θ_{XB} is small, enhancing pulsation transfer.   Conversely, we would expect an IMF transverse to the solar wind ($\theta_{XB} = 90°$) to remove turbulence entirely from the subsolar shock, suppressing pulsation transfer.   These configurations are depicted in Fig. 12 (GREENSTADT, 1972).

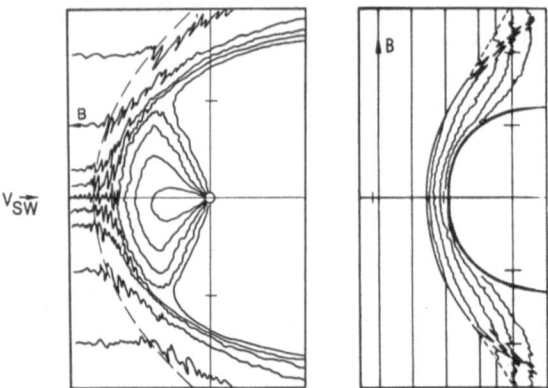

Fig. 12.   Field orientations most favorable, left, and unfavorable, right, to delivery of
quasi-parallel bow shock turbulence to the dayside magnetopause.

3.2   Critique

3.2.1   Surface-wave generation alone

The Kelvin-Helmholtz mechanism would presumably operate on any initial perturbation at the magnetopause, whether a density, velocity, or pressure fluctuation.   The
initial perturbation would not necessarily have to be a wave, and such perturbations are
assumed to be present in the magnetosheath most of the time.   At this writing, no concept
has been advanced that would explain the cone angle, or any other IMF directional effect
as a strictly secondary contribution.   Indeed, the threshold criterion for K-H instability
contains the factor $B_1^2 \cos^2 \psi_1$ where B_1 is the magnetosheath field strength at the boundary
and ψ_1 is the angle between the field and the solar wind flow at the boundary (BOLLER and
STOLOV, 1973).   This factor is *least* favorable to the instability when the flow and field are
aligned, and they would be aligned everywhere on the magnetopause when cone angle
$\theta_{xB} = 0$.

Certainly the most prominent characteristic of the IMF direction known to influence the magnetosphere is its north-south component, but investigation has failed
to disclose any correlation between interplanetary B_Z and daytime pulsation signals.

Added to the difficulty that the K-H model has so far provided no clue to inverse cone
angle dependence, there is a more serious obstacle to this mechanism acting alone; namely,
the heavy bias of pulsation activity toward morning, even dawn, occurrence, while the solar
wind velocity and its induced surface waves should be essentially symmetric on east
and west sides of the magnetopause.   This difficulty is not quite so serious if we divide the
pulsation band into short and long periods.   The low-latitude measurements of RAO
(1979) have characterized Pc 4 and Pc 5 as peaking in occurrence both after dawn and after
dusk.   On the other hand, Pc 3 are well-known as a daytime phenomenon with maximal
occurrence before noon.   In the context of the models we are discussing, the observations
suggest that Pc 4 and Pc 5 are associated with the flanks, Pc 3 with the subsolar sections
of the magnetosphere.   The former are therefore compatible with K-H generation,
the latter with quasi-parallel.   The results of WOLFE *et al.* (1979) and WOLFE (1980) correlating V_{SW} most strongly with long period activity are also consistent with these associa-

tions.   On balance, the evidence favors a strong, if not exclusive, role for K-H at the longer periods.

3.2.2   Quasi-parallel wave excitation alone

We have postulated that waves of large amplitude, formed in the subsolar bow shock when the cone angle is less than 50° and 70° (depending on how loosely we define the subsolar region), are propagated and convected to the daytime magnetopause, where they transfer to the magnetosphere and through the magnetosphere to the ground.

Figure 13 illustrates the relative influence of two principal factors governing the delivery of large amplitude, quasi-parallel shock-structured waves through the magnetosheath to the magnetopause.   These factors are: Solar wind velocity (speed and direction) and Alfvén wave velocity.

As the streamlines in the figure show, only solar wind passing through the subsolar region of the shock impinges directly on, or comes very close to, the magnetopause. Waves convected with the wind on the flank of the shock can reach the dayside magnetopause only if they cross the sheath in less time than the plasma carrying them takes to sweep them downstream.   The small pairs of arrows indicate, at the intersection of each pair, the relative velocities of the solar wind in its local direction of flow, and the Alfvén wave velocity toward the magnetopause.   The flow and wave velocities have been estimated from the computations of SPREITER and ALKSNE (1969) for the IMF in the ecliptic (the plane of the figure) at 45° to the sun-earth line.   The small, circled numbers in the magnetosheath in the figure indicate approximately the local ratio C_A/V_{SW} (i.e., M_A^{-1})

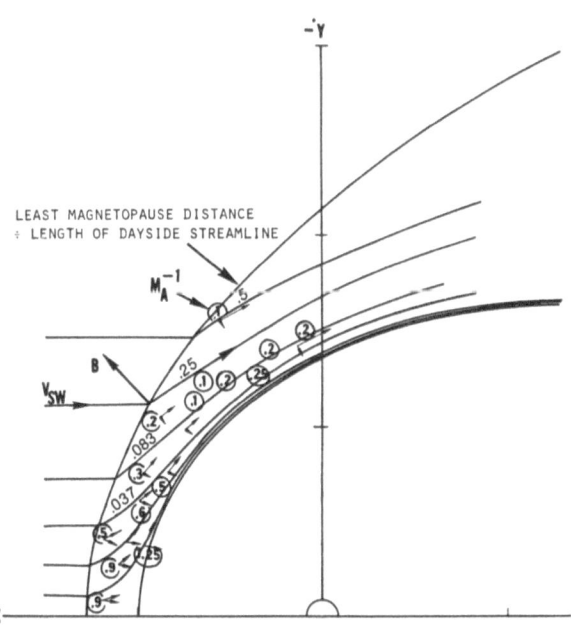

Fig. 13.   Solar wind streamlines in the magnetosheath and wind-wave speed parameters
   for 45° IMF.

where C_A is the Alfvén speed (M_A is the Alfvén Mach number).

It is clear in Fig. 13 that not only does plasma entering the flank of the magnetosheath avoid the magnetopause, but waves originating in the far flank plasma are so slow relative to V_{SW} that they are unlikely to propagate to the magnetopause until far downstream. In contrast, subsolar plasma not only streams toward the magnetopause, but its parameteric properties support faster wave propagation than at the flank, making delivery of waves to the magnetopause highly probable. The figure suggests that up to some angle from the subsolar point, quasi-parallel waves could cross layers of solar wind and reach the dayside magnetopause, while beyond that angle they could not. The numbers on the streamlines just inside the nominal shock outline are rough estimates of the shortest distance from the streamline to the magnetopause divided by the length of the corresponding streamline segment up to the terminator (the $-Y$ axis). These numbers are a crude guide to the ratio of C_A to V_{SW} needed to pass shock oscillations to the magnetopause. If the distance ratio is less than the velocity ratio (the circled numbers), the waves may reach the magnetopause; otherwise, they will not. The figure suggests, then, that waves reaching the magnetopause must originate in the shock no more than about 45° or so from the subsolar point, when the IMF is at 45°. Such a requirement is compatible with the model of shock origin, since a 45° IMF will cause quasi-parallel structure to prevail in the subsolar region. The structure, however, will be transitional around the subsolar point, where $\theta_{Bn} \approx 45°$, and increasingly oscillatory as θ_{Bn} becomes smaller. If we think of the IMF and the cross section of the figure as lying in the solar ecliptic plane, then the largest structural waves of the parallel configuration will occur on the morning side. This is significant for the model because, although not shown in Fig. 13, wave ratios on the afternoon side are higher, making wave transfer to the afternoon magnetopause easier than to the morning side if waves are present in the subsolar or post noon shock. The implication then is that the bulk of wave excitation must occur on average as a result of large amplitude q-parallel structural waves in the *forenoon* sector, west of the subsolar point itself.

On average over an extended time interval, the morning sector would be preferentially excited, while waves moving through the magnetosphere would encounter conditions favorable to resonance at certain field lines, generally around and outside the plasmapause. The exact lines would depend on the wave period and the state of the magnetosphere.

There is an interesting implication of associating flow-aligned field (i.e., $\theta_{XB}=0$) with pulsation origin. Recall that the best correlation of pulsation activity with θ_{XB} has been found for short periods, and that the inverse relation $T_{pc} \propto B^{-1}$ has involved just these shorter periods, less than 60 to 100 seconds (Fig. 1). We recognize this inverse proportionality as the same one defining the proton cyclotron period. More specifically, we can write $T=160/B=67/.42B$, making T the cyclotron period of protons in a field $B'=.42 B$. If we consult the diagram of Spreiter and Alksne (1969), we see it is precisely in a field-aligned flow that the field near the subsolar magnetopause is reduced to a fraction of its level in the upstream solar wind. Thus, the influence of low cone angle in producing pulsations in the Pc 3 and lower half of the Pc 4 bands is consistent with the way such periods would vary with B if the pulsations resulted from waves at the cyclotron period near the subsolar magnetopause.

The velocity effect can be accounted for as a secondary phenomenon in any of three ways: First, high velocity is associated with Mach number which makes a more complex shock structure and moves the nominal shock closer to the magnetopause shortening the distance over which waves would need to be convected through the magnetosheath. Second, a high velocity would suggest that some part of the q-parallel wave energy that might ordinarily move upstream in a slower wind would be blown downstream into the sheath, adding its measure to the energy available for excitation of the magnetopause. Third, high solar wind speed is associated with a perturbed magnetosphere in which, among other things, the resonance lines may move in just such a way as to bring enlarged amplitudes to the rather limited latitudes of sensors whose data have contributed to the established correlations. Any one of the above mechanisms or any combination of them could produce an increase in pulsation amplitude with velocity without invoking surface-wave amplification by Kelvin-Helmholtz instability.

3.2.3 Joint action

The joint action of wave excitation and wave transfer has been modeled as one in which the Kelvin-Helmholtz instability amplifies, as surface waves, oscillations delivered to the magnetopause from quasi-parallel structure in the subsolar section of the bow shock. The cone angle is clearly important in determining whether, and to what extent, the flux tubes directed at the magnetopause are occupied by large-amplitude waves. The speed is important in determining the efficiency and degree of amplification according to classical surface wave theory. The joint action appears to be manifested in a way consistent with a "gating" effect in which a favorable angle switches in magnetosheath signals, permitting the continuously-available velocity mechanism to function. Alternatively, and perhaps more realistically, unfavorable angles switch out the perturbations usually available to the velocity mechanism. We reason as follows: The statistical studies have demonstrated that the speed correlation is the more reliable and sharply defined relationship. Solar wind speed is in general a rather slowly changing parameter; in addition, those low speeds for which no surface wave amplification would be expected are very low and comparatively unusual, corresponding only to the least disturbed solar and interplanetary conditions. We would expect, therefore, that pulsations would be produced almost continuously, with infrequent lapses preceding the onsets of fast streams. This is substantially what Fig. 8 shows.

The statistical studies have also demonstrated that the cone angle correlation is comparatively weak and subject to wide scatter, but is improved by narrowing the time intervals over which the cone angle and pulsations are examined. Unlike the solar wind speed, or, for that matter, the IMF magnitude, the IMF direction, including the cone angle, is subject to appreciable short term variation. Like the speed, however, the cone angle is more likely to be favorable than unfavorable to conditions thought responsible for pulsations. A study of bow shock structural statistics (GREENSTADT, 1973) based on IMF orientations compiled by FAIRFIELD and NESS (1967) from IMP 2 data put the quasi-parallel oscillations in the subsolar region of the shock 40 to 50% of the time for the cited compilation.

To remove quasi-parallel structure entirely from the subsolar region probably requires a cone angle of more than 70°, but no model has been worked out yet, or observations assembled, for estimating the extent to which the subsolar region of the shock should be involved in wave structure as a function of angle. The amplitude of waves as a function of local normal angle is unknown, and no exact definition of what constitutes the subsolar region for this purpose has been determined. It is clear, however, that there is far more likelihood than unlikelihood that most of the time *some* q-parallel oscillation will be present between the *X*-axis and, say, the second streamline in Fig. 13. Moreover, *some* waves should be able to reach the magnetopause from the shock beyond the second streamline, as already argued. On balance, we expect waves to be ordinarily present in the subsolar region, and we expect their enlargement or disappearance to occur irregularly as transient events of duration thirty minutes or less. We envision the surface wave mechanism operating almost continuously, generating a background or base level of pulsation signal at the prevailing solar wind speed, subject to transient increases and decreases governed by swings in cone angle. The longer the time unit, or interval, over which data are averaged, compiled, or characterized, the weaker the angle effect will appear with respect to the speed effect.

Finally, we acknowledge that the observational evidence of Rao (1979) and Wolfe *et al.* (1980) suggest that at the longest periods the K-H mechanism may operate independently of θ_{XB}. So far, the data supporting this picture have involved averaging over hourly intervals, so the angle effect would be weakened anyway, and a conclusive test with short intervals has yet to be conducted.

4. Summary

A train of differing, but related, observational results over an interval of more than a decade has demonstrated that an important part, if not the total, of the natural ULF magnetic spectrum in the period range 10 to 300 sec (Pc 3, 4, 5) consists of magnetospheric signals whose amplitudes, periods, and occurrences are controlled by the state of the solar wind. In addition, two models have been postulated that would account for the signals as waves in the magnetosphere actually generated at and beyond the magnetopause. An attempt at interpretation of the evidence as a whole results in the following picture: The solar wind passing through the subsolar bow shock and magnetosheath flows into and around a stagnation region outside the subsolar magnetopause and then flows along the magnetopause, picking up speed along the flanks, both east-west and north-south. Toward the terminator meridian, the wind has accelerated enough to exceed the threshold for Kelvin-Helmholtz instability, and surface waves of periods roughly 50 to 300 sec are amplified and transmitted into the magnetosphere. The higher the solar wind speed, the more effective the process and the larger the amplitudes of the pulsations inside the magnetosphere.

At the same time, large amplitude waves generated as part of the structure of the quasi-parallel bow shock predominate in the forenoon magnetosheath under average conditions. The waves reach the morning magnetopause by propagation and convection and penetrate or stimulate the magnetopause, sending waves into and through the dayside magnetosphere. The waves in the sheath are largest and should by themselves influence the magnetopause most strongly in the subsolar region when the IMF rotates toward alignment with the solar wind flow. When this happens, the waves in the inner sheath have local cyclotron periods, roughly between 10 and 100 sec, which are related inversely to the strength of the IMF. Pulsations stimulated in the magnetosphere then share these same periods. When variation of the IMF removes the quasi-parallel waves from the subsolar or forenoon magnetosheath, stimulation of the dayside magnetosphere is halted.

In addition to improving the growth of surface waves, higher solar wind speed may also improve the efficiency of quasi-parallel wave convection to, or transmittal through, the magnetosphere, while quasi-parallel waves may also reach the terminator region, providing a source of the perturbation needed to trigger the Kelvin-Helmholtz instability there. Thus, there may be joint, combined, or overlapping effects of plasma and field parameters on pulsation occurrence.

5. Observational Program

The natural outcome of the preceding discussion is to suggest a number of projects around which further investigation could center with promise of important results. Table 3 lists a set of tasks, or research programs, that can be done with IMS and other data to advance our understanding of daytime pulsations. The first five items are suitable for individual initiative; the last is an enterprise of a cooperative nature, indeed of international interest, on which we elaborate further.

Table 3. Observational tasks.

1. Establish the fluctuation pattern in the sheath.
2. Investigate q-parallel wave amplitude vs. θ_{Bn}, especially near subsolar point.
3. Study quantitatively the wave content of q-parallel structure vs. mach number.
4. Examine the velocity and cone angle correlations of long-period (Pc 5) pulsations with short sampling intervals.
5. Seek cases of connection between presence or absence of magnetosheath perturbations, magnetopause waves, and pulsation activity.
6. Define an index of global daytime activity for $10 < T < 300$ sec, free of local resonance, f^{-x} spectral skewing, and plasma trough variability, and apply the index to correlations with solar wind parameters.

5.1 Development of a world wide pulsation index

Control of magnetospheric pulsations by the solar wind implies that it may eventually be possible to use pulsations to monitor solar wind parameters. At the current time, however, this is not possible, for several reasons.

First, there may be more than one source of energy in the midperiod bands, requiring separation of that portion attributable to external sources. Second, there is no uniform, worldwide system for detecting, recording, and processing pulsation signals to produce measurements independent of extraneous local time effects. Third, there is no developed technique for suppressing the dominating effects of local resonances to produce measurements independent of latitude, geomagnetic activity, and accidental station circumstances. These difficulties need to be overcome if the concept of a pulsation index is to advance to a serious quantitative tool for monitoring the earth's plasma environment.

We believe the possibilities for removing the foregoing obstacles are realizable. First, it may be possible to resolve Pc signal sources by the kind of data processing described by OLSON and SAMSON (1979), who have devised an economical means of distinguishing signals by polarization states.   Second, reasonably consistent digital recording of pulsation bands at high resolution has been put into practice at many locations (i.e., longitudes).   It remains to complete a global network and establish a cooperative effort at standardizing a data reduction, compression, and display format.   Third, there exist already several chains of stations distributed in latitude at constant magnetic longitude.   It remains to compare records from any one chain over a comprehensive sequence of conditions, process the data uniformly, and devise a scheme for normalizing the pulsation signal levels to eliminate domination by the capricious appearance of resonances at different observations at different times.

We are hopeful that a standardized pulsation measure, or a set of indices, will be created in the foreseeable future and employed to improve significantly our understanding and our exploration of the solar wind control of geomagnetic pulsations.

REFERENCES

ARTHUR, C. W. and R. L. MCPHERRON, Interplanetary magnetic field conditions associated with synchronous orbit observations of Pc 3 magnetic pulsations, *J. Geophys. Res.*, **82**, 5138–5142, 1977.

BOLLER, B. R. and H. L. STOLOV, Explorer 18 study of the stability of the magnetopause using a Kelvin-Helmholz stability criterion, *J. Geophys. Res.*, **78**, 8078–8086, 1973.

BOL'SHAKOVA, O. V. and V. A. TROITSKAYA, Relation of the interplanetary magnetic field direction to the system of stable oscillations, *Dokl. Akad. Nauk SSSR*, **180**, 4–6, 1968.

CHEN, L. and A. HASEGAWA, A theory of long-period magnetic pulsations, 1, Steady state excitation of field line resonance, *J. Geophys. Res.*, **79**, 1024–1032, 1974.

D'ANGELO, N., Are Pc 2–4 micropulsations of extramagnetospheric origin?, *Geomagn. Aeron.*, **15**, 746–747, 1975.

FAIRFIELD, D. H. and N. F. NESS, Magnetic field measurements with the 1MP 2 Satellite, *J. Geophys. Res.*, **72**, 2379–2402, 1967.

GREENSTADT, E. W., Field-determined oscillations in the magnetosheath as possible source of medium-period, daytime micropulsations, in *Proceedings of Conference on Solar Terrestrial Relations*, 515 pp., Univ. of Calgary, April 1972.

GREENSTADT, E. W., Statistics of bow shock nonuniformity, *J. Geophys. Res.*, **78**, 2331–2336, 1973.

GREENSTADT, E. W., The solar wind and magnetospheric waves, Magnetospheric Study 1979, in *Proceedings of International Workshop on Selected Topics of Magnetospheric Physics*, p. 160, Japanese IMS Committee, Tokyo, 1979.

GREENSTADT, E. W. and R. W. FREDRICKS, Shock system in collisionless space plasmas, in *Solar System Plasma Physics*, Vol. 3, edited by C. F. Kennel, L. G. Lanzerotti, and E. N. Parker, p. 4, North-Holland, New York, 1979.

GREENSTADT, E. W. and J. V. OLSON, Pc 3, 4 activity and interplanetary field orientation, *J. Geophys. Res.*, **81**, 5911–5920, 1976.

GREENSTADT, E. W. and J. V. OLSON, A contribution to ULF activity in the Pc 3–4 range correlated with IMF radial orientation, *J. Geophys. Res.*, **82**, 4991–4996, 1977.

GREENSTADT, E. W. and J. V. OLSON, Geomagnetic pulsation signals and hourly distributions of IMF orientations, *J. Geophys. Res.*, **84**, 1493–1498, 1979.

GREENSTADT, E. W., H. J. SINGER, C. T. RUSSELL, and J. V. OLSON, IMF orientation, solar wind velocity, and Pc 3–4 signals: A joint distribution, *J. Geophys. Res.*, **84**, 527–532, 1979a.

GREENSTADT, E. W., J. V. OLSON, P. D. LOEWEN, H. J. SINGER, and C. T. RUSSELL, Correlation of Pc 3, 4, and 5 activity with solar wind speed, *J. Geophys. Res.*, **84**, 6694–6696, 1979b.

GRINGAUZ, K. I., E. K. SOLOMATINA, V. A. TROITSKAYA, and R. V. SHCHEPETNOV, Variations of solar wind flux observed by several spacecraft and related pulsations of the earth's electromagnetic field, *J. Geophys. Res.*, **76**, 1065–1069, 1971.

GUL'ELMI, A. V., Diagnostics of the magnetosphere and interplanetary medium by means of pulsations, *Space Sci. Rev.*, **16**, 331–344, 1974.

GUL'ELMI, A. V. and O. V. BOL'SHAKOVA, Diagnostics of the interplanetary magnetic field from ground-based data on Pc 2–4 micropulsations, *Geomagn. Aeron.*, **13**, 535–537, 1973.

GUL'ELMI, A. V., T. A. PLYASOVA-BAKUNINA, and R. V. SHCHEPETNOV, Relation between the period of geomagnetic pulsations Pc 3, 4 and the parameters of the interplanetary medium at the earth's orbit, *Geomagn. Aeron.*, **13**, 331–333, 1973.

KAUFMANN, R. L., J. T. HORNG, and A. WOLFE, Large-amplitude hydromagnetic waves in the inner magnetosheath, *J. Geophys. Res.*, **75**, 4666–4676, 1970.

KOVNER, M. S., V. V. LEBEDEV, T. A. PLYASOVA-BAKUNINA, and V. A. TROITSKAYA, On the generation of low-frequency waves in the solar wind in the front of the bow shock, *Planet. Space Sci.*, **24**, 261–267, 1976.

LEBOEUF, J. N., T. TAJIMA, C. F. KENNEL, and J. M. DAWSON, Global magnetohydrodynamic simulation of the two-dimensional magnetosphere, in *Quantitative Modeling of Magnetsospheric Processes, Geophysical Monograph*, Vol. 21, edited by W. P. Olson, 536 pp., Am. Geophys. Union, Washington D.C., 1979.

LEBOEUF, J. N., T. TAJIMA, C. F. KENNEL, and J. M. DAWSON, Global simulations of the three-dimensional magnetosphere, preprint PPG-450, UCLA Cent. for Plasma Phys. and Fusion Eng., Jan. 1980.

NOURRY, G. R., Interplanetary magnetic field, solar wind and geomagnetic micropulsations, Thesis, Univ. of British Columbia, Dept. of Geophysics and Astronomy, 1976.

OLSON, J. V. and J. C. SAMSON, On the detection of the polarization states of Pc micropulsations, *Geophys. Res. Lett.*, **6**, 413–416, 1979.

RAO, D. R. K., Morphological features of the low-latitude pulsations, Magnetospheric Study 1979, in *Proceedings of International Workshop on Selected Topics of Magnetospheric Physics*, p. 150, Japanese IMS Committee, 1979.

RUSSELL, C. T. and B. K. FLEMING, Magnetic pulsations as a probe of the interplanetary magnetic field: A test of the Borok *B*-index, *J. Geophys. Res.*, **81**, 5882–5886, 1976.

SAITO, T., K. YUMOTO, K. TAKAHASHI, T. TAMURA, and T. SAKURAI, Solar wind control of Pc 3, Magnetospheric Study 1979, in *Proceedings of International Workshop on Selected Topics of Magnetospheric Physics*, p. 155, Japanese IMS Committee, Tokyo, 1979.

SINGER, H. J., C. T. RUSSELL, M. G. KIVELSON, E. W. GREENSTADT, and J. V. OLSON, Evidence for the control of Pc 3, 4 magnetic pulsations by the solar wind velocity, *Geophys. Res. Lett.*, **4**, 377–379, 1977.

SOUTHWOOD, D. J., The hydromagnetic stability of the magnetospheric boundary, *Planet. Space Sci.*, **16**, 587–605, 1968.

SOUTHWOOD, D. J., Some features of field line resonances in the magnetosphere, *Planet. Space Sci.*, **22**, 483–491, 1974.

SPREITER, J. R. and A. Y. ALKSNE, Plasma flow around the magnetosphere, *Rev. Geophys.*, **1**, 11–50, 1969.

TAKAHASHI, K., R. L. MCPHERRON, E. W. GREENSTADT, and C. W. ARTHUR, Factors controlling the occurrence of Pc 3 magnetic pulsations at synchronous orbit, *J. Geophys. Res.*, submitted, 1980.

TROITSKAYA, V. A., T. A. PLYASOVA-BAKUNINA, and A. V. GUL'ELMI, Relationship between Pc 2–4 pulsations and the interplanetary magnetic field, *Dokl. Akad. Nauk. SSSR*, **197**, 1312, 1971.

TROITSKAYA, V. A., R. V. SHCHEPETNOV, and A. V. GUL'ELMI, Sudden disappearance of type Pc 2–4 geomagnetic pulsations, *Geomagn. Aeron., USSR*, **9**, 294–296, 1969.

VERÓ, J., Geomagnetic pulsations and parameters of the interplanetary medium, Magnetospheric Study 1979, in *Proceedings of International Workshop on Selected Topics of Magnetospheric Physics*, p. 177, Japanese IMS Committee, Tokyo, 1979.

VERÓ, J. and L. HOLLÓ, Connections between interplanetary magnetic field and geomagnetic pulsations, *J. Atmos. Terr. Phys.*, **40**, 857–865, 1978.

VINOGRADOV, P. A. and V. A. PARKHOMOV, MHD waves in the solar wind—a possible source of geomagnetic Pc 3 pulsations, *Geomagn. Aeron., USSR*, **15**, 109–112, 1974.

WEBB, D. and D. ORR, Geomagnetic pulsations (5–50 mHz) and the interplanetary magnetic field, *J. Geophys. Res.*, **81**, 5941–5947, 1976.

WEBB, D., L. J. LANZEROTTI, and D. ORR, Hydromagnetic wave observations at large longitudinal separations, *J. Geophys. Res.*, **82**, 3329–3335, 1977.

WOLFE, A., Dependence of midlatitude hydromagnetic energy spectra on solar wind speed and interplanetary field direction, *J. Geophys. Res.*, **85**, 5977–5982, 1980.

WOLFE, A., L. J. LANZEROTTI, and C. G. MACLENNAN, Dependence of hydromagnetic energy spectra on interplanetary parameters, *EOS*, **60**, 360, 1979.

WOLFE, A., L. J. LANZEROTTI, and C. G. MACLENNAN, Dependence of hydromagnetic energy spectra on solar wind velocity and interplanetary magnetic field direction, *J. Geophys. Res.*, **85**, 114–118, 1980.

Pulsation Structure in the Ionosphere Derived from Auroral Radar Data

A. D. M. WALKER* and R. A. GREENWALD**

*Department of Physics, University of Natal, Durban, South Africa
**Applied Physics Laboratory, Johns Hopkins University, Maryland, U.S.A.

(Received June 28, 1980)

It has recently become apparent that auroral radars are powerful tools for the investigation of ULF pulsations in the Pc 5 range. Auroral radar observations of pulsation effects in the ionosphere are reviewed. The STARE radar system is described and its measurements of the pulsation electric field in the ionosphere outlined. It has been established that this electric field is entirely consistent with the field of a hydromagnetic field-line resonance. The rotation of the pulsation magnetic field through 90° by the ionosphere is confirmed. Statistical studies of occurrence show some consistency with the behaviour expected for solar-wind driven instabilities on the magnetopause. The use of STARE data to deduce equatorial plasma density, and height integrated ionospheric conductivities is also discussed. These measurements have been used to estimate energy input into the magnetosphere.

1. Introduction

The understanding of the nature of Pc 5 geomagnetic pulsations has advanced substantially during recent years. This has been due on the one hand to the introduction of new techniques of measurement such as the availability of large magnetometer chains (e.g. SAMSON and ROSTOKER, 1972; SAMSON et al., 1971), coordinated satellite measurements (e.g. HUGHES et al., 1977, 1978, 1979), and auroral radars (UNWIN and KNOX, 1971; MCDIARMID and MCNAMARA, 1972; WALKER et al., 1978, 1979), and on the other hand to strong theoretical advances (SOUTHWOOD, 1974; CHEN and HASEGAWA, 1974; HUGHES, 1974; HUGHES and SOUTHWOOD, 1976a, b). The purpose of this paper is to review one of these experimental techniques—the use of auroral radars—which is particularly suited to the observation of Pc 5 pulsations.

First we discuss the nature of auroral radar reflections from E-region irregularities and review early observations of Pc 5 pulsation phenomena with such instruments. We then discuss the STARE (Scandinavian twin auroral radar experiment) system which provides an effective means of measuring the ionospheric electric field associated with pulsations, with high spatial resolution (20 km × 20 km) over an area of about 200,000 km². The comparison of these measurements with theory provides strong evidence that these phenomena are magnetic field line resonances of the type described by SOUTHWOOD (1974) and CHEN and HASEGAWA (1974). Support is also provided for the prediction that the ionosphere rotates the magnetic field of the pulsation through 90° (HUGHES and SOUTHWOOD, 1976a, b). Statistics of occurrence are presented and provide some support for the Kelvin-Helmholtz mechanism. Finally the use of such measurements for deducing ionospheric and magnetospheric parameters and the importance of pulsations as an energy transport mechanism between solar wind and ionosphere are discussed.

2. Radar Auroral Observations Associated with Pc 5 Pulsations

Radar aurora consists of irregularities in the auroral E region of the ionosphere between about 100 and 120 km altitude (Unwin, 1966). These irregularities are field-aligned and produce radar backscatter when the radar beam is very nearly at right angles to the earth's magnetic field, B. This aspect sensitivity is very critical and the intensity of backscatter is down by 20 dB from the maximum when the beam is only 3° from the strict normal to B.

The irregularities arise because, in this region, ions are collision dominated, while electrons are not. Thus, when there is an electric field, E, transverse to B, the electrons drift with an $E \times B$ drift, while ions undergo both Pedersen and Hall type motions. If the relative velocity of ions and electrons is sufficiently large two-stream and gradient-drift instabilities occur (Buneman, 1963; Farley, 1963; Rogister and D'Angelo, 1970). Non-linear effects limit the growth and allow the structure to cascade down to shorter wavelengths (Sudan et al., 1973; Greenwald, 1974).

The theory predicts that the irregularities drift with the electron velocity. This is supported by observational evidence (Ecklund et al., 1977; Cahill et al., 1978). The irregularities have been observed with wavelengths varying from tens of meters to a few millimeters and thus cause backscatter for frequencies from tens to hundreds of megahertz.

2.1 Early observations of Pc 5 by auroral radar

The observations discussed in this section are summarized in Table 1. The type of observations is typified by Fig. 1 which is taken from one of the earliest papers (Kaneda et al., 1964). It represents a so-called range-time-intensity (RTI plot). Geomagnetic latitude is plotted vertically and universal time horizontally. The shaded areas represent times when, and latitudes from which, backscatter was obtained. Also shown are magnetometer traces which demonstrate that the fluctuations in the radar aurora have the same period as Pc 5 pulsations. An obvious feature of the diagram is the poleward drift of the regions from which backscatter occurs. Later workers (Unwin and Knox, 1971; McDiarmid and McNamara, 1972, 1973) also measured the Doppler velocity of the irregularities.

Some of the results obtained by these workers can be summarized as follows:

i) On an RTI plot, at a fixed range, the intensity of backscatter varies with a period equal to that of the associated pulsation. The regions from which backscatter occurs appear to drift polewards.

ii) Unwin and Knox (1971) and McDiarmid and McNamara (1972, 1973) found that the radial patch velocity (which is different from the Doppler velocity) was linearly related to the pulsation frequency. The Doppler velocity varied in an inverse V shaped pattern on each pulsation and at times changed sign.

Early attempts to explain the results assumed that the scattering regions were caused

Table 1

Authors	Date	Type of observation
TIURI	1960	(Unpublished report cited by Keys, 1960)
KANEDA, KOKUBUN, OGUTI, and NAGATA	1964	Intensity of backscatter. Instruments located in Alaska.
Keys	1965	Intensity of backscatter at 55 MHz. Instrument located in New Zealand.
BROOKS	1967	Same instrument as Keys.
UNWIN and KNOX	1971	Intensity of backscatter and Doppler velocity of irregularities at 488 MHz. Instrument located in Canada.
McDIARMID and McNAMARA	1972, 1973	Intensity of backscatter and Doppler velocity of irregularities at 48 MHz, 488 MHz, and 944 MHz. Instruments located in Canada.

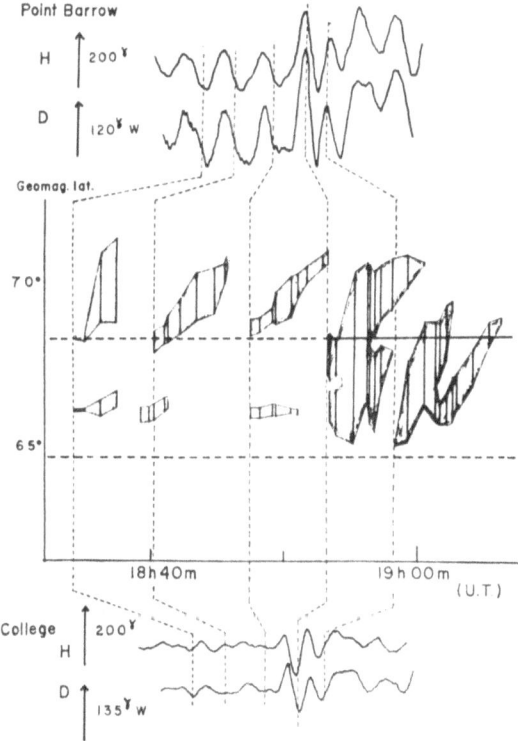

Fig. 1. An early example of the relationship between an auroral radar range-time record and Pc 5 geomagnetic activity (reproduced, with permission, from KANEDA et al., 1964).

by particle precipitation (KANEDA et al., 1964; KEYS, 1965). Later UNWIN and KNOX (1971) suggested the electric field associated with a standing hydromagnetic wave as the source. McDIARMID and McNAMARA (1973) suggested that the radar aurora and Pc 5 pulsations were associated through a plasma instability.

2.2 The STARE radar

The STARE radar (GREENWALD et al., 1978) is a significant advance on previous auroral radar systems. It consists of two pulsed radars at Trondheim, Norway and Hankasalmi, Finland. Each consists of a broad beam transmitting antenna, a multiple narrow beam receiving antenna, and a shipping container holding transmitter, receivers and on-line processing equipment. The radars measure intensity and Doppler velocity of irregularities at 50 different ranges along each of 16 beams, with a temporal resolution of 20 s or 60 s under normal operating conditions. In Fig. 2 the 8 most frequently used beams of each system are shown. They have a large common area in which, at each point, the Doppler velocities along two different directions can be found and used to yield the electron drift velocity and hence the electric field. Examples of the drift velocities and backscatter intensities during a Pc 5 event are shown in Fig. 3 which is taken from WALKER *et al.* (1979).

2.3 *STARE observations of pulsations*

Considerable progress has been made in understanding the nature of pulsations by the use of the STARE system (WALKER *et al.*, 1978, 1979). Much of the remainder of the

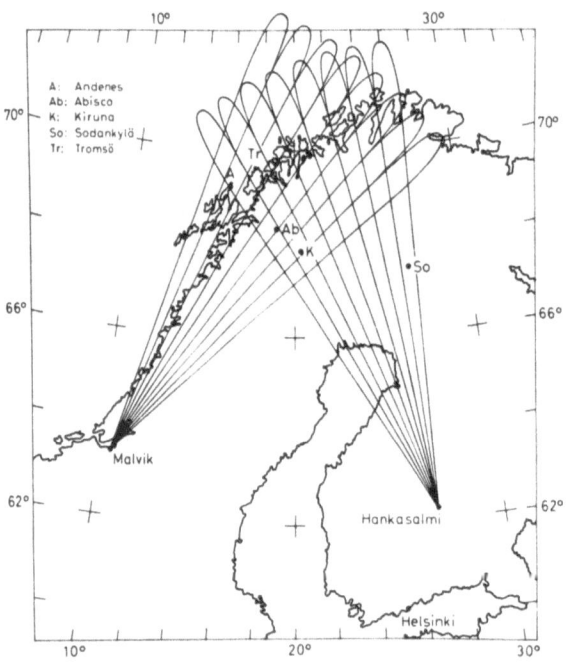

Fig. 2. Map of Scandinavia in geographic coordinates showing the 8 central lobes of each of the Stare radars. The overlapping region of the lobes covers 230,000 km. The instruments are located at Hankasalmi and Malvik (Trondheim) (from WALKER *et al.*, 1979).

Fig. 3. Backscatter intensity and irregularity drift velocity averaged over 20 s at 4 times during an afternoon event on Day 303, 1977. Each panel consists of two smaller plots giving the intensity of backscatter as a function of geographic latitude and longitude for each of the 2 radars. The larger plot represents the drift velocity of the irregularities. Each spot is a grid point at which the velocity has been measured. The line segments represent velocity vectors (from WALKER *et al.*, 1979).

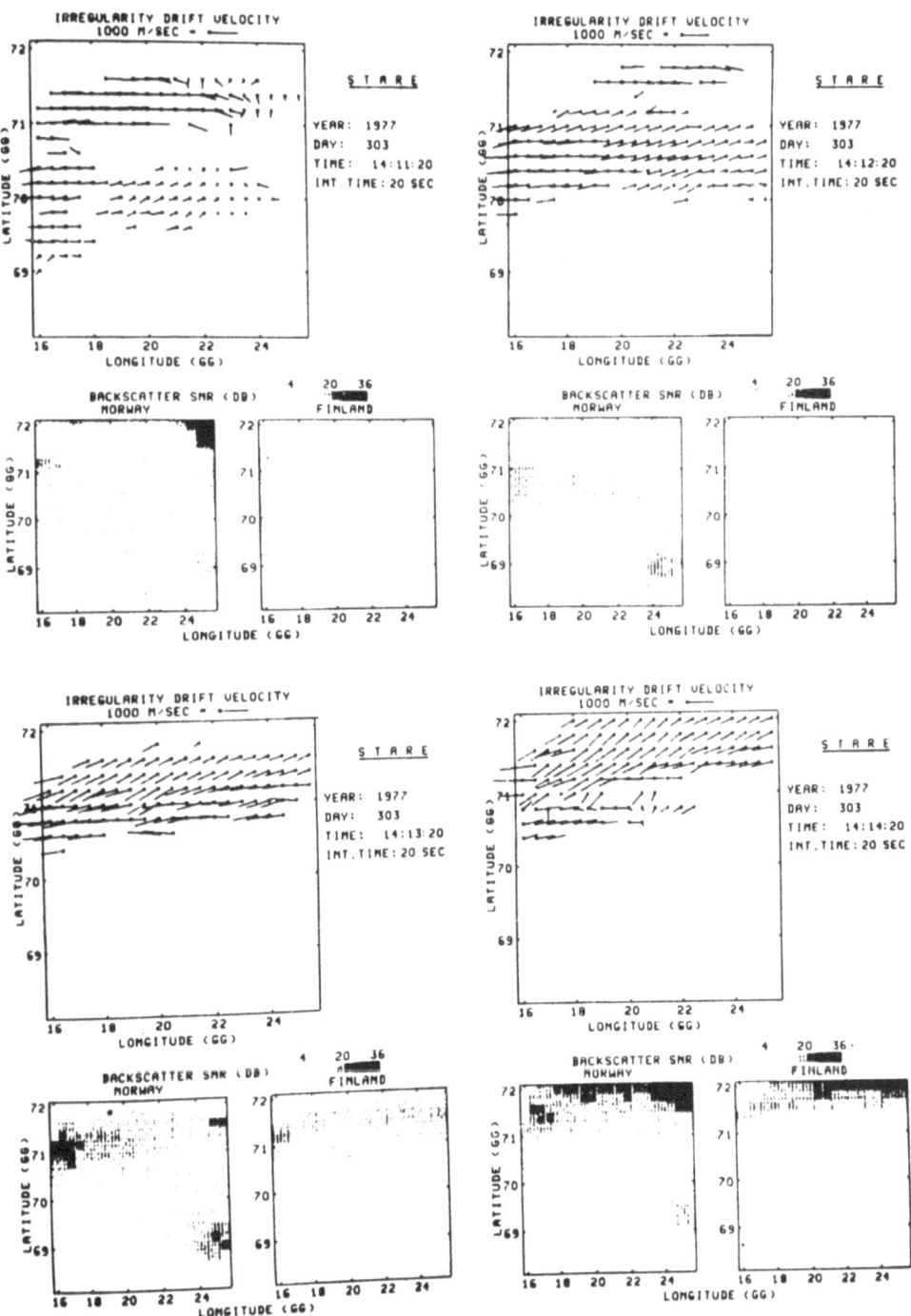

Fig. 3

review is concerned with this work. Figure 3 is divided into 4 panels. Each panel is a so-called STARE plot and consists of a large and two small plots. The small plots show the backscatter intensity for Finland and Norway as a function of position within the STARE field of view. The intensity is represented on a grey scale. The large plot represents the drift velocity of the irregularities. Each spot is a grid point at which the resultant velocity has been found. Each line segment represents a velocity vector directed away from the grid point. Because the electric field is $-V \times B$ and B is downward and nearly vertical, the electric field can be found by rotating the vectors 90° clockwise and scaling 1 km s^{-1} to 50 mV m^{-1}. Each panel in Fig. 3 represents the result of a 20 s integration during a Pc 5 event. The panels are spaced one minute apart. Intermediate data exists but has been omitted for conciseness. It can be seen that, for this event, there are alternate bands of eastward and westward drift velocity, the bands themselves drifting polewards. This is, of course, consistent with the behaviour observed by Mc-DIARMID and McNAMARA (1973).

This behaviour is shown in another way in Fig. 4. Here the longitude range has been restricted to lie between 18° and 20°. At each latitude within this range the velocity has been averaged over longitude. These data have then been plotted as a function of latitude and U.T. to show the time behaviour. The poleward drift of the alternate bands of eastward and westward drift velocity is again apparent in this diagram.

3. Hydromagnetic Resonances

3.1 SCH theory

Work done using magnetometer chains, especially that of SAMSON et al. (1971) and SAMSON and ROSTOKER (1972) showed that the polarization of a Pc 5 geomagnetic pulsation is a function of local time and of station latitude. This led SOUTHWOOD (1974) and CHEN and HASEGAWA (1974) independently to propose a theory of hydromagnetic resonances to explain the magnetometer data. This theory (hereafter SCH theory after the initials of its originators) was successful in explaining the magnetometer results and made further predictions which could not be verified by magnetometers because of an inherent lack of spatial resolution due to the effect of the earth-ionosphere cavity on the signal.

The basis of SCH theory is as follows: It can easily be shown that, in a cylindrically symmetrical geometry, magnetic shells can oscillate toroidally with cylindrical symmetry independently of adjacent shells. The plasma motion is thus everywhere in a direction perpendicular to the magnetic meridian plane. The frequency of this toroidal oscillation is determined by the plasma density and by the length of a field line. In an ideal case the ionosphere is a node of the electric field but if the ionosphere has a finite conductivity the oscillation is damped. The natural frequency of oscillation is a monotonic decreasing function of L except in regions where the plasma density changes rapidly, such as the plasmapause. It is assumed, in SCH theory, that, as illustrated in Fig. 5, the solar wind sets up monochromatic waves on the magnetopause through the Kelvin-Helmholtz instability. These waves drive an oscillatory motion in the magnetic meridian, decaying with distance from the magnetopause, but penetrating deep into the magnetosphere. Because

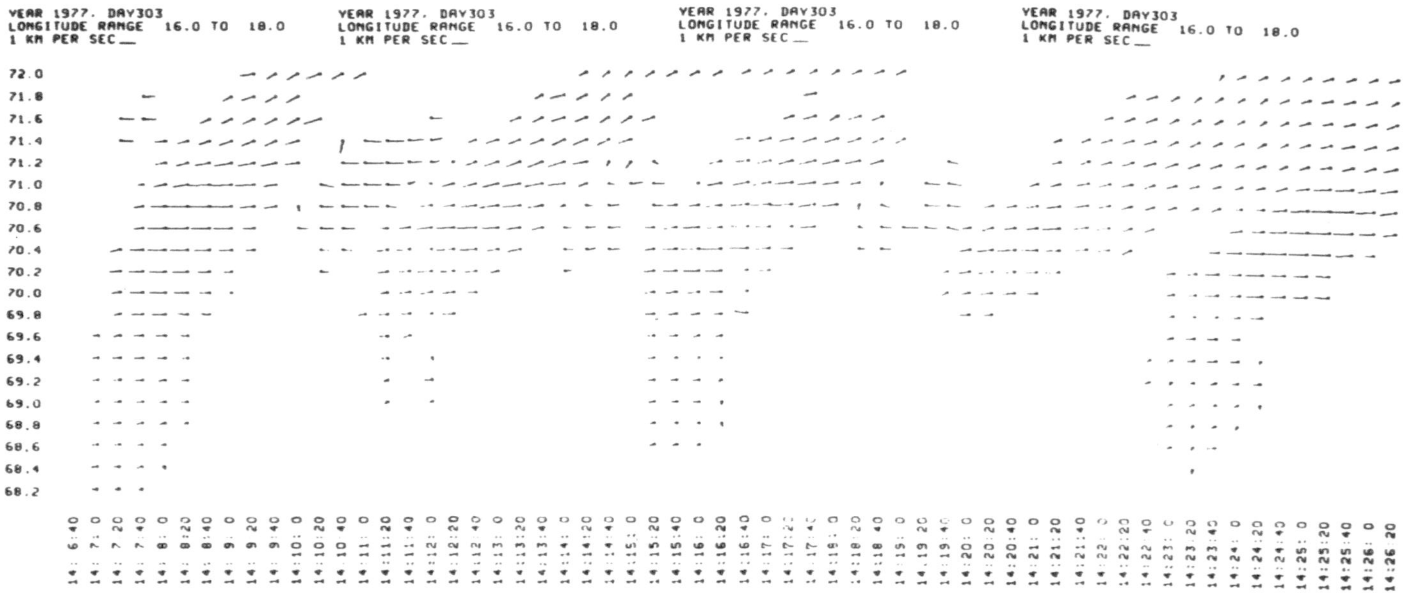

Fig. 4. Mean irregularity velocity in the geographic longitude range 16.0° for the event shown in Fig. 3 as a function of geographic latitude (vertical axis) and universal time (horizontal axis) (from WALKER et al., 1979).

of the dipole geometry this motion is coupled to the toroidal motion. If the natural frequency of toroidal oscillation matches that of the Kelvin-Helmholtz wave, resonance takes place. The resulting motion of the plasma in the equatorial plane is shown in the figure and maps down to the ionosphere. WALKER (1980) has used the theory to model the structure of pulsation fields and some of his results are shown in Fig. 6. In particular the theory predicts a ∼180° shift of phase of the north-south component of the electric field in the ionosphere, and a resonance peak in its amplitude.

3.2   Analysis of data

STARE data has been analysed by WALKER et al. (1979) and fits these predictions very

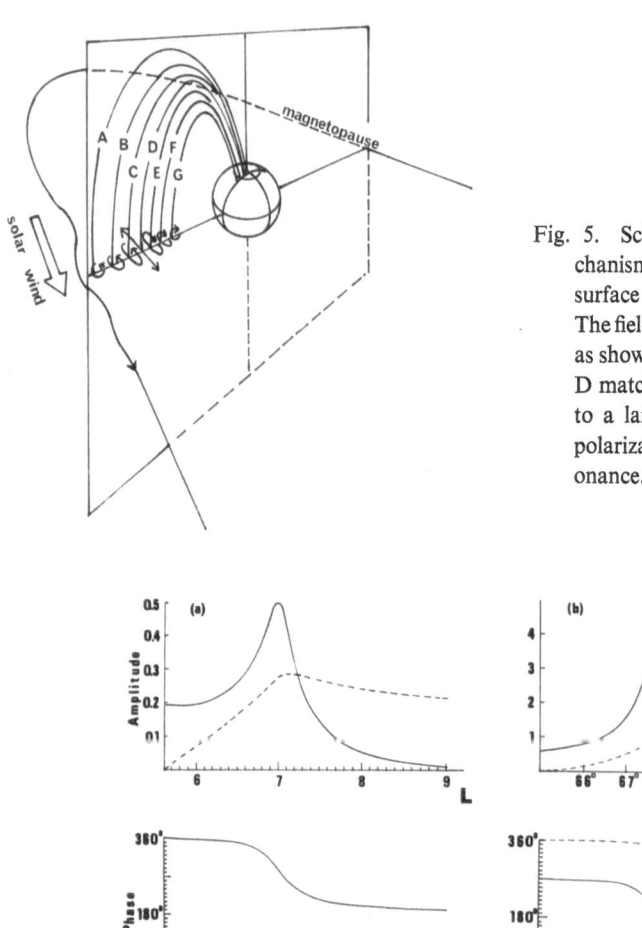

Fig. 5. Schematic diagram of SCH mechanism. The solar wind causes a surface wave on the magnetopause. The field lines A, B, C, D, E, F, G move as shown. The toroidal frequency of D matches the wave frequency leading to a large toroidal component. The polarization changes across the resonance.

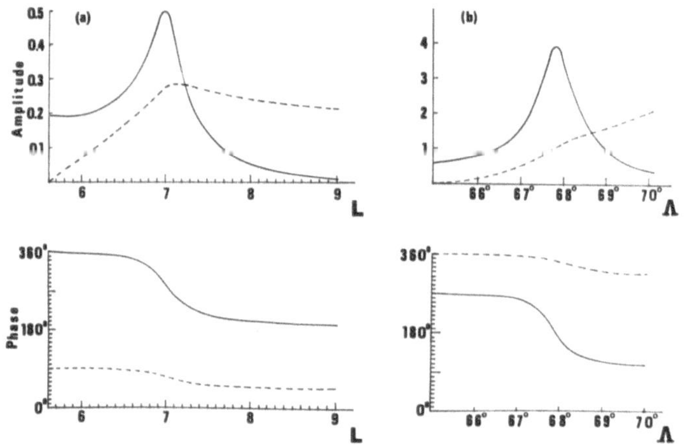

Fig. 6. Amplitude and phase of the electric fields in the meridian (full line) and perpendicular to the meridian (dashed line):   (a) In the equatorial plane expressed as a function of L. (b) At ground level expressed as a function of Λ, the geomagnetic latitude. $k_1/k_0 = 0.12$, $q = 0$, $m = 5$, $L_B = 5.6$, $L_R = 6.2$ (from WALKER (1980), where notation is described).

well.   In Fig. 7 a series of Fourier amplitude spectra of the drift velocity are shown for the event shown in Fig. 4.   Each panel is plotted to the same scale, represents a different latitude, and shows that there is a peak in the spectrum corresponding to the resonance

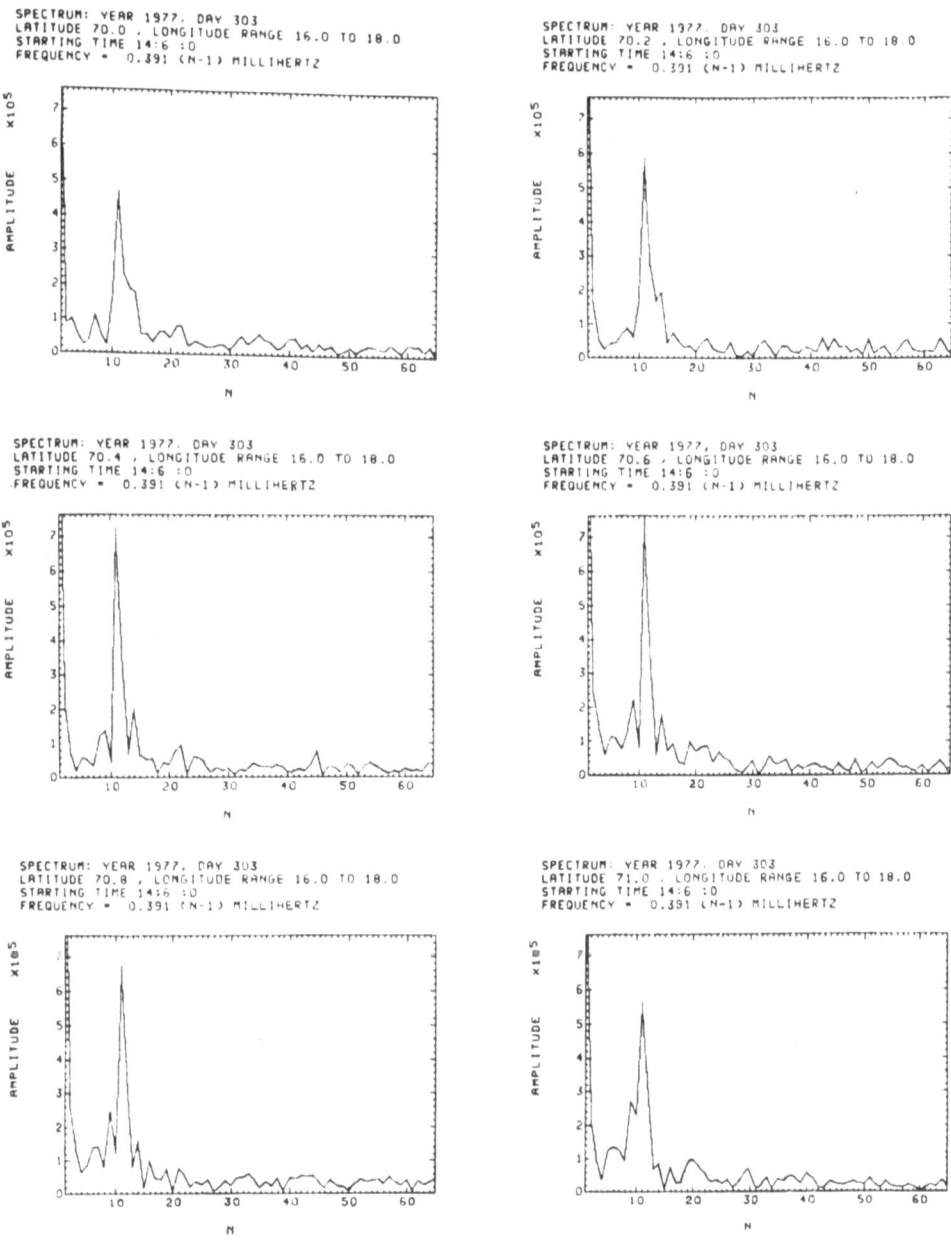

Fig. 7. Amplitude spectra of the eastward velocity component for the event of Fig. 3 at a number of different latitudes.   The vertical scale is the same for each panel (from WALKER *et al.*, 1979).

frequency. The amplitude of this peak is maximum at the latitude corresponding to resonance. Both amplitude and phase of this frequency component are plotted as functions of latitude in Fig. 8 and clearly behave qualitatively exactly as expected by SCH theory. WALKER *et al.* (1979) have considered a number of events in which this is so.

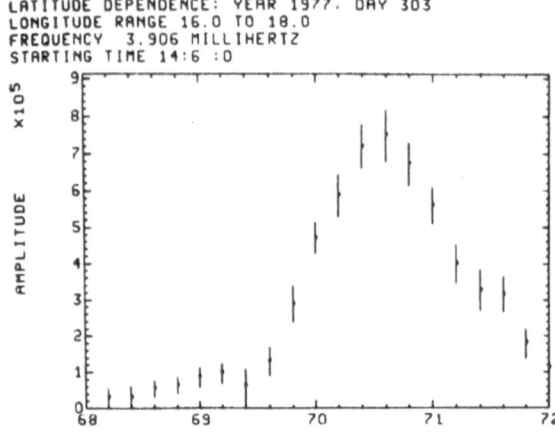

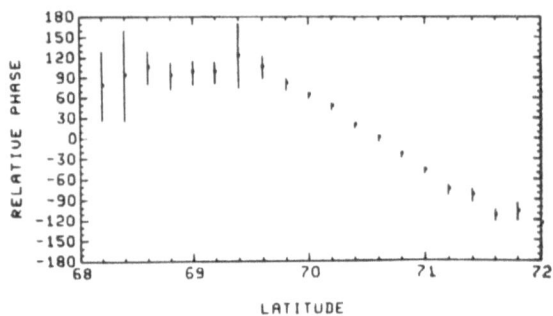

Fig. 8. Upper panel: Amplitude of the 3.906 mHz component of the spectrum of the eastward velocity component for the event of Fig. 3 averaged over the longitude range 16.0°–18.0° as a function of latitude. Lower panel: Corresponding phase, relative to that at latitude 70.6° (from WALKER *et al.*, 1979).

4.   The Effect of the Ionosphere

Various authors (NISHIDA, 1964; INOUE, 1973; HUGHES, 1974) have predicted that the effect of the ionosphere is to rotate the horizontal component of the magnetic field of a long period pulsation (but not the electric field) through 90°. This is because the Pedersen current in the ionosphere screens the magnetic field of the pulsation from the ground, while the field on the ground arises from the associated Hall current.

The STARE data have also been compared with magnetometer data (WALKER *et al.*, 1978, 1979). An example of the magnetic field on the ground predicted by STARE is compared with magnetometer data in Fig. 9. This has been done assuming that the field on the ground arises entirely from the Hall current. The good agreement gives support to the idea of ionospheric rotation described above.

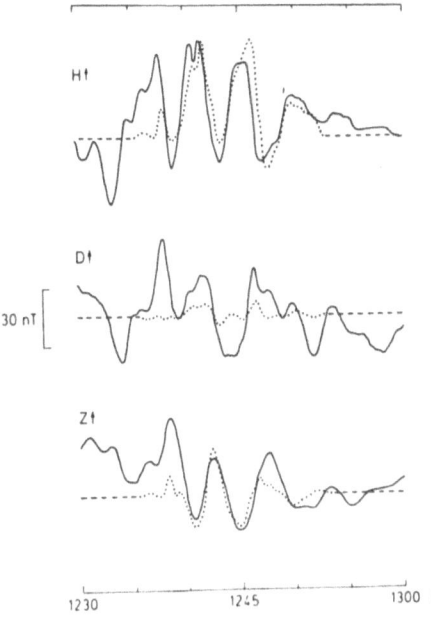

Fig. 9. Comparison of magnetometer data and field predicted on the ground using the Biot-Savart Law for an afternoon event on day 331, 1977. Solid line shows magnetometer data. Dotted line shows field predicted by Stare data. Dashed line shows Stare data missing because the disturbance was below threshold (from WALKER et al., 1978).

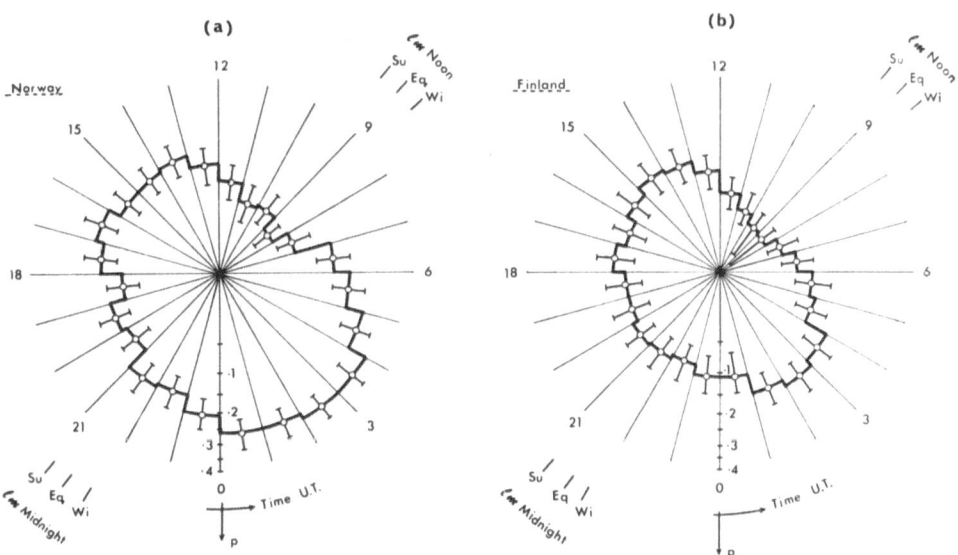

Fig. 10. Circular histogram of occurrence frequency of Pc 5 period events observed by Stare for the year November 1977 to October 1978 inclusive. The number of events is proportional to the area of each segment (from WALKER and GREENWALD, 1980).

5. Statistics of Occurrence

While the work of WALKER et al. (1978, 1979) provides clear evidence for the essential correctness of SCH theory there is no direct evidence of what drives the resonance. No phase change in the east-west direction can be observed within the 400 km range of STARE, and thus there is no guarantee that the resonance is associated with a wave propagating tailwards as would be the case if the Kelvin-Helmholtz mechanism were operative. One indirect way of getting additional evidence is to consider the statistics of occurrence of auroral radar events. This has been done by WALKER and GREENWALD (1980). It is to be expected that the waves on the magnetopause should be maximum on the flanks of the magnetosphere where the shear across the boundary is largest.

WALKER and GREENWALD (1980) have analysed one year of STARE RTI data for a single beam for each station. They define an event as having occurred if, within any observing hour, they can see clear spatial evidence in the data of a poleward drifting band, and if there are at least three cycles of oscillation in the hour. Figure 10 summarises their data. While it shows a clear minimum at local magnetic midday, there is little evidence for a minimum at midnight. If the data is classified according to magnetic activity, however, a different picture emerges. Figure 11 shows the same data separated out for

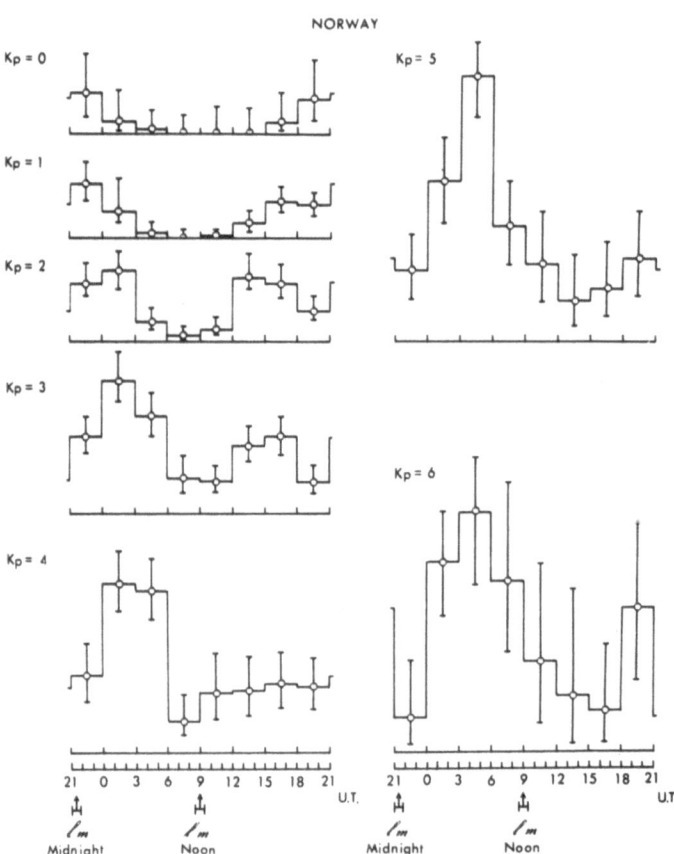

Fig. 11. Occurrence frequency of Pc 5 period events observed by Stare for the year November 1977 to October 1978 inclusive. Classified according to the value of K_p.

different values of K_p for one of the stations. This shows clear morning and afternoon maxima for low values of K_p. At higher values of K_p, however, there is a strong morning peak and the positions of the minima are different. It appears that the data are consistent with the Kelvin-Helmholtz picture at low K_p, but, at higher K_p, there is less clarity.

6. STARE Pulsation Observations as Diagnostics

With the clearer understanding of the nature of radar aurora associated with geomagnetic pulsations it is possible to use such observations to deduce several magnetospheric and ionospheric parameters.

From the measured frequency and an observation of the resonance latitude WALKER et al. (1979) found values of the plasma number density outside the plasmasphere, on the assumption that the dominant ion is H^+. They found values varying between 15 and 50 particles per cm^3 for different events. If, as now seems likely there are a substantial number of heavy ions present, this can be interpreted as a mass density between 2.5×10^{-23} and 8.5×10^{-23} g cm^{-3}. This information is also obtainable from standard RTI data and it is to be hoped that workers in the field will make use of the technique. All that is necessary is to establish the range at which the resonance peak occurs and hence the geomagnetic latitude of resonance as well as the frequency. The plasma density can then be found from the theory of ORR and MATTHEW (1971).

Another possibility is to obtain a value of the height-integrated Pedersen conductivity by measuring the width of the resonance peak. A particular example of the technique is given by WALKER (1980). Here the theory of hydromagnetic resonance is used to fit a theoretical curve to a set of experimental points. The adjustable parameter determining the width of the resonance peak is the height-integrated ionospheric Pedersen conductivity. An example of such a fit is shown in Fig. 12. It is estimated that the precision of the method is about 15–20%. The method depends on the assumption of a dipole magnetic field which may introduce errors. Nevertheless application of the technique to the events discussed by WALKER et al. (1979) leads to height-integrated conductivities of 2.5–4.5 Ω^{-1}. These are closely in agreement with values obtained by other workers using different techniques for the same time of day in the auroral region.

WALKER et al. (1979) also compared ground based magnetometer observations with the ground magnetic field predicted by the STARE data. They obtained a good fit and were able from this to make rough estimates of the height-integrated Hall conductivity. This method is limited by the finite field of view of STARE as assumptions must be made about the behaviour of the current outside the field of view. It leads to plausible values of about 10 Ω^{-1} for the height-integrated Hall conductivity.

Finally the theory presented by WALKER (1980) holds out hope that, by comparing the phase of pulsations observed by satellite-borne instruments and STARE, it will be possible to identify the field line (or at least the magnetic shell) on which the satellite is located. All that is necessary is to identify the locus on the ground at which the signal is 90° out of phase with the signal observed on the satellite. This locus represents the line where the magnetic shell, on which the satellite is located, cuts the ionosphere. This could allow the testing of geomagnetic field models which are at present used for the

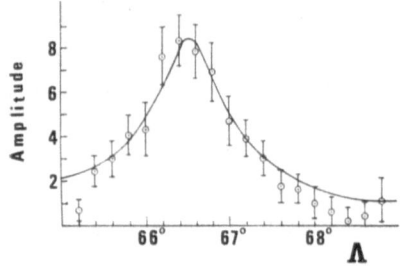

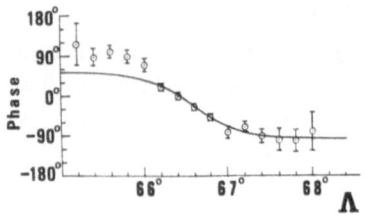

Fig. 12.   Fit of data for Feb. 2 1977 to theory. Upper panel: amplitude. Lower panel: phase.

purpose of identifying the field line on which the satellite is located.

7.   Energetics of Pc 5 Pulsations

STARE measurements of the electric field and height-integrated ionospheric Pedersen conductivity allow the rate of energy deposition into the ionosphere to be calculated. GREENWALD and WALKER (1980) have done this for several of the events described by WALKER *et al.* (1979). They find that the rate of energy deposition can peak as high as 5.8 mW m^{-2} for one of these events. They estimate the area over which the pulsation is significant as 1.2×10^{12} cm^2 and hence a rate of energy dissipation 7×10^9 W through joule heating (7×10^{-6} erg s^{-1}). This compares with $2 \times 10^{10} - 10^{11}$W ($2 \times 10^{17} - 10^{18}$ erg s^{-1}) for the joule heating associated with a substorm and is an appreciable fraction of the total energy dissipated by a small substorm. They also compute the total field-aligned currents associated with the hydromagnetic resonance. These can be found either from the component of the curl of the pulsation magnetic field along $\boldsymbol{B}_0$, or from the divergence of the ionospheric current. This field-aligned current can be as large as several micro-amps per square meter-large enough to excite topside current-driven instabilities. They thus predict that it may be possible to see auroral arcs drifting poleward under the control of the pulsation. It is possible that some of the arcs seen by OGUTI *et al.* (1979) may be of this type. Workers should be careful not to assume that, if pulsations are seen in conjunction with arcs, then the auroral particles give rise to the pulsation. The reverse may well be true.

8.   Conclusions

Auroral radar observations of the electric fields associated with Pc 5 geomagnetic pulsations are capable of substantially better spatial resolution than ground-based measurements. This has nothing to do with the inherent nature of the instruments used, but is a characteristic of the phenomenon itself. The spatial scale of the pulsation is so small that information about it is smeared out by the earth-ionosphere cavity. The nature of at least one class of Pc 5 pulsation is now clearly identified as a hydromagnetic resonance phenomenon. The hydromagnetic structure is a self-consistent system of electric and magnetic fields, and currents. The name "geomagnetic pulsation" thus seems something of a misnomer when applied to the phenomenon, stressing, as it does, the magnetic aspect. "ULF pulsation" would be a better description of such a phenomenon. It is clear that future progress is going to come from the application of a combination of techniques such as ground-based magnetometer chains, satellite magnetic and electric field measurements, radar techniques, and, possibly, ground-based and balloon electric field measurements. Auroral radars can be expected to make a substantial contribution to this progress. Such radars should ideally be of the STARE type. However single beam radars can also make a substantial contribution to the observation of Pc 5 pulsations.

We can summarize our conclusions thus:

1)   Pc 5 pulsations are hydromagnetic field line resonances, peaking over a narrow band of latitudes.

2)   There is nothing in the direct observations which is inconsistent with a solar wind driven wave on the magnetopause as the source of the phenomenon. There is, however, no direct evidence for this. The statistical evidence hints that there may be more than one type of driving source. This is the most important outstanding problem requiring solution.

3)   Routine radar observations of Pc 5 pulsations could be used to monitor magnetospheric plasma densities and height-intergrated Pedersen ionospheric conductivities.

4)   The energy input into the ionosphere by a large pulsation is considerable, and should be included in the total energy transfer from solar wind to ionosphere. If it is assumed that the pulsation arises from a solar wind driven wave on the magnetopause, then it provides a direct pipeline for energy transfer from solar wind to magnetosphere.

5)   The field-aligned currents associated with the hydromagnetic resonance are large enough to produce topside current instabilities, and it is predicted that poleward moving auroral arcs should be associated with the pulsation.

A. D. M. Walker is grateful for support from the South African Council for Scientific and Industrial Research.

REFERENCES

BROOKS, D., Radio auroral echoes associated with sudden commencements and their possible use in measurements of magnetospheric ion densities, *J. Atmos. Terr. Phys.*, **29**, 589–597, 1967.

BUNEMAN, O., Excitation of field aligned sound waves by electron streams, *Phys. Rev. Lett.*, **10**, 285–287, 1963.

CAHILL, L., R. A. GREENWALD, and E. NIELSEN, Auroral radar and rocket double-probe observations of the electric field across the Harang discontinuity, *Geophys. Res. Lett.*, **5**, 687–690, 1978.

CHEN, L. and A. HASEGAWA, A theory of long-period magnetic pulsations, 1, Steady state excitation of field line resonance, *J. Geophys. Res.*, **79**, 1024–1032, 1974.

ECKLUND, W. L., B. B. BALSLEY, and D. A. CARTER, A preliminary comparison of F-region plasma drifts and *E* region irregularity drifts in the auroral zone, *J. Geophys. Res.*, **82**, 195–197, 1977.

FARLEY, D. T., A plasma instability resulting in field-aligned irregularities in the ionosphere, *J. Geophys. Res.*, **68**, 6083–6097, 1963.

GREENWALD, R. A., Diffuse radar aurora and the gradient drift instability, *J. Geophys. Res.*, **79**, 4807–4810, 1974.

GREENWALD, R. A. and A. D. M. WALKER, Energetics of long period hydromagnetic waves, *Geophys. Res. Lett.*, 1980 (in press).

GREENWALD, R. A., W. WEISS, E. NIELSEN, and N. P. THOMSON, Stare: A new radar auroral backscatter experiment in northern Scandinavia, *Radio Sci.*, **13**, 1021–1039, 1978.

HUGHES, W. J., The effect of the atmosphere and ionosphere on long period magnetospheric micro-pulsations, *Planet. Space Sci.*, **22**, 1157–1172, 1974.

HUGHES, W. J. and D. J. SOUTHWOOD, The screening of micropulsation signals by the atmosphere and ionosphere, *J. Geophys. Res.*, **81**, 3234–3240, 1976a.

HUGHES, W. J. and D. J. SOUTHWOOD, An illustration of modification of pulsation structure by the iono-sphere, *J. Geophys. Res.*, **81**, 3241–3247, 1976b.

HUGHES, W. J., R. M. MCPHERRON, and C. T. RUSSELL, Multiple satellite observations of pulsation re-sonance structure, *J. Geophys. Res.*, **82**, 492–498, 1977.

HUGHES, W. J., R. L. MCPHERRON, and J. N. BARFIELD, Geomagnetic pulsations observed simultaneously on three geo-stationary satellites, *J. Geophys. Res.*, **83**, 1109–1116, 1978.

HUGHES, W. J., R. L. MCPHERRON, J. N. BARFIELD, and B. H. MANK, A compressional Pc 4 pulsation ob-served by three satellites in geo-stationary orbit near local midnight, *Planet. Space Sci.*, **27**, 821–840, 1979.

INOUE, Y., Wave polarization of geomagnetic pulsations observed in high latitudes on the earth's surface, *J. Geophys. Res.*, **78**, 2959–2976, 1973.

KANEDA, E., S. KOKUBUN, T. OGUTI, and T. NAGATA, Auroral radar echoes associated with Pc 5, *Rep. Ionos. Space. Res. Japan*, **18**, 165–172, 1964.

KEYS, J. G., Pulsating auroral radar echoes and their possible hydromagnetic association, *J. Atmos. Terr. Phys.*, **27**, 385–393, 1965.

MCDIARMID, D. R. and A. G. MCNAMARA, Periodically varying radar aurora, *Ann. Geophys.*, **28**, 433–441, 1972.

MCDIARMID, D. R. and A. G. MCNAMARA, Radio aurora, storm sudden commencements, and hydro-magnetic waves, *Can. J. Phys.*, **51**, 1261–1265, 1973.

NISHIDA, A., Ionospheric screening effect and storm sudden commencements, *J. Geophys. Res.*, **69**, 1861–1874, 1964.

OGUTI, T., K. HAYASHI, S. KOKUBUN, K. TSURUDA, T. WATANABE, and R. E. HORITA, Auroral and magnetic pulsations, in *Magnetospheric Study 1979, Proceedings of the International Workshop on Selected Topics of Magnetospheric Physics*, pp. 111–115, Japanese IMS Committee, Tokyo, 1979.

ORR, D. and J. A. D. MATHEW, The variation of geomagnetic pulsation periods with latitude and plasma-pause, *Planet. Space Sci.*, **19**, 897–905, 1971.

ROGISTER, A. and N. D'ANGELO, Type II irregularities in the equatorial electrojet, *J. Geophys. Res.*, **75**, 3879–3887, 1970.

SAMSON, J. C. and G. ROSTOKER, Latitude dependent characteristics of high-latitude Pc 4 and Pc 5 micro-pulsations, *J. Geophys. Res.*, **77**, 6133–6144, 1972.

SAMSON, J. C., J. A. JACOBS, and G. ROSTOKER, Latitude dependent characteristics of long period geo-magnetic micropulsations, *J. Geophys. Res.*, **76**, 3675–3683, 1971.

SOUTHWOOD, D. J., Some features of field line resonances in the magnetosphere, *Planet. Space Sci.*, **22**, 483–491, 1974.

SUDAN, R. N., J. AKINRIMISI, and D. T. FARLEY, Generation of small scale irregularities in the equatorial electrojet, *J. Geophys. Res.*, **78**, 240–248, 1973.

UNWIN, R. S., The morphology of the v.h.f. radio aurora at sunspot maximum, 2, The behaviour of different echo types, *J. Atmos. Terr. Phys.*, **28**, 1183–1194, 1966.

UNWIN, R. S. and F. B. KNOX, Radio aurora and electric fields, *Radio Sci.* **6**, 1061–1077, 1971.

WALKER, A. D. M., Modelling of Pc 5 pulsation structure in the magnetosphere, *Planet. Space Sci.*, **28**, 213–223, 1980.

WALKER, A. D. M. and R. A. GREENWALD, Statistics of occurrence of hydromagnetic oscillations in the Pc 5 range observed by the Stare auroral radar, *Planet. Space Sci.*, 1980 (in press).

WALKER, A. D. M., R. A. GREENWALD, W. F. STUART, and C. A. GREEN, Resonance region of a Pc 5 micropulsation examined by a dual auroral radar system, *Nature*, **273**, 646–649, 1978.

WALKER, A. D. M., R. A. GREENWALD, W. F. STUART, and C. A. GREEN, Stare auroral radar observations of Pc 5 geomagnetic pulsations, *J. Geophys. Res.*, **84**, 3373–3388, 1979.

Damping and Coupling of Long-Period Hydromagnetic Waves by the Ionosphere

F. B. KNOX and W. ALLAN

Physics and Engineering Laboratory, Private Bag, Lower Hutt, New Zealand

(Received June 28, 1980)

The theory that long-period geomagnetic pulsations are due to hydromagnetic waves resonating in a magnetosphere bounded by an anisotropically conducting ionosphere, insulating atmosphere, and conducting earth, is reviewed. The history of the subject is outlined, covering the effects of the ionospheric boundary on: pulsations seen at ground level and in the ionosphere (ionospheric screening, 90° rotation of polarisation, attenuation of short wavelengths, mode coupling via Hall conductance); and wave systems above the ionosphere (wave-boundary impedance matching, damping, phase variation along the geomagnetic field, and allowed resonances). Coupling of non-axisymmetric modes within the magnetosphere is also mentioned, and a summary of the present picture given, with suggestions for future work.

1. Introduction

In this review of geomagnetic pulsation theory, 'long-period' means in the range 20 to 600 sec, and covers the periods of Pc 3, 4, 5, and Pi 2 pulsations.

It is currently believed long-period pulsations are due to hydromagnetic waves in the magnetosphere, with wavelengths comparable to the dimensions of the magnetosphere. Thus hydromagnetic resonance must be an important factor in the theory of such waves. The development and present state of the resonance theory as such will be reviewed, with particular reference to the effect of the ionosphere as a wave boundary. Excitation of the waves, a subject in which progress has also been made, will not be discussed.

2. History

Hydromagnetic wave theory was first applied to the magnetosphere by DUNGEY (1954), who restricted consideration to axisymmetric waves, with particular reference to the

---

List of symbols: r, θ, φ, spherical polar coordinate system (aligned with axis of, and centred on earth's dipole magnetic field); ν, μ, φ, dipole coordinate system (RADOSKI, 1967) (aligned with axis of, and centred on earth's dipole magnetic field); h_ν, h_μ, h_φ, metric functions of the dipole coordinate system; m, wave mode number $(1, 2, 3, 4, \ldots)$; N, S, superscripts referring respectively to northern and southern ionospheres; 0, 1, superscripts referring respectively to just above and just below a sheet ionosphere; μ_0, permeability of free space; R_E, earth radius; L, McIlwain parameter; B_0, geomagnetic induction; a_p, Pedersen conductance (height-integrated conductivity); a_H, Hall conductance (height-integrated conductivity); A, Alfvén speed; ζ_I, wave admittance evaluated at the ionosphere; b, wave magnetic induction; E, wave electric field; I, current per unit length; u, plasma velocity; $\dot{D}$, displacement current; S, Poynting vector; ω_0, angular frequency at resonance; γ, damping decrement; k, $(\omega_0 + i\gamma)/A$.

situation in a dipole field. He introduced the two basic solutions of the axisymmetric wave equation, later known as the toroidal and poloidal modes. The toroidal is also known as the transverse, torsional or Alfvén mode, and the poloidal as the compressional or fast mode.

Even in an axisymmetric geomagnetic field the equations, in general, must have coupled non-axisymmetric toroidal and poloidal solutions, but these have not so far been expressed analytically. However analytical solutions of the uncoupled axisymmetric equations have been obtained in certain cases. These are believed to give valid wave periods, and, in recent development, to describe approximately the structure in meridian cross-section.

As shown in Fig. 1 resonant toroidal solutions may to first order be visualised as single dipole shells oscillating azimuthally (east-west) about the dipole axis, while the poloidal solutions fill the whole space available and oscillate radially.

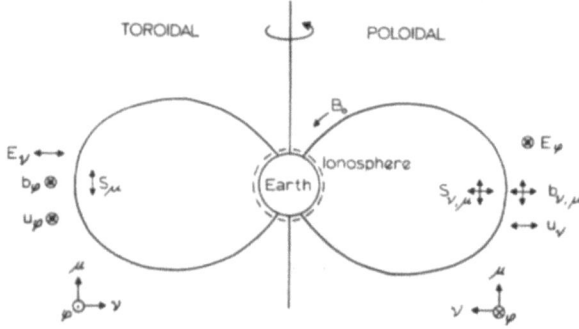

Fig. 1. Axisymmetric toroidal and poloidal modes.

HRUSKA (1968) pointed out that the toroidal single shell solutions violate one of the assumptions used in deriving the equations themselves, viz. that there is a finite rate of change of electric field perpendicular to the shell. Nevertheless he considered the deduced resonant periods to be valid.

In this early work Dungey assumed the earth-side boundary to be an infinitely con-ducting solid, anchoring the geomagnetic field lines, so that no plasma velocity was allowed at this boundary. Hence nodes of wave motion (and electric field) could be assumed at the earth's surface. However further work by DUNGEY (1958, 1963) revealed the importance of the insulating atmosphere in largely isolating the wave field from the ground, allowing the ionosphere to play a significant part in bounding the wave.

Thus far the effect of an *anisotropically* conducting ionosphere had been ignored. However, in the course of a study of magnetic storm sudden commencements, NISHIDA (1964) showed that the magnetic induction of a hydromagnetic disturbance, propagating through an anisotropically conducting ionosphere, appeared to rotate through 90°.

VAN'YAN and ABRAMOV (1969a, b) also obtained 90° rotation of the magnetic induc-tion from one side to the other of an anisotropic ionosphere, for a simple hydromagnetic wave model, in which a transverse electric dipole source drove a wave with geomagnetic field-aligned currents flowing in and out of a sheet ionosphere. This early result was confirmed, for more realistic wave configurations and ionospheric conductivity versus height profiles, by INOUE (1973) and HUGHES (1974).

In addition to the above modification of the wave field as it passes through the ionosphere, HUGHES and SOUTHWOOD (1976a, b) pointed out that there is also attenuation between the ionosphere and the ground, for geometrical reasons concerned with how the ionospheric height relates to the horizontal scale of the wave. This effect is particularly important for the toroidal mode which is localized in latitude.

The effect of the ionosphere on the wave in the magnetosphere was treated by VAN'YAN et al. (1970), VAN'YAN and KOZHEVNIKOV (1974), HUGHES (1974), and HUGHES and SOUTH-WOOD (1976a, b). They found that only the Pedersen component of the anisotropic conductance in relation to the wave admittance was important in determining the configuration of a toroidal wave far above the ionosphere. What the Hall component of the ionospheric conductance does is to couple the toroidal wave to an evanescent poloidal wave which attenuates rapidly with height.

MAL'TSEV and LEONT'YEV (1977) derived an explicit expression for ionospheric damping of a hydromagnetic wave, by modelling the wave as a two-conductor transmission line terminated at one end by the ionosphere. The damping expression of Mal'tsev and Leont'yev agrees approximately with the numerical calculations of NEWTON et al. (1978) for two extreme forms of magnetospheric standing wave—the axisymmetric toroidal, and the so-called quasi-transverse (or guided) poloidal mode. The latter mode is a non-compressional oscillation confined to a meridian plane.

The numerical solutions of Newton et al. for finite ionospheric conductivities, symmetric about the equator, revealed that typical nightside conductivities should give e-fold damping of the wave in about three periods, but dayside conductivities should only give this amount of damping in 10 to 20 periods. Newton et al. also showed that the phase of the electric and magnetic components of the wave vary continuously along the geomagnetic field. This contrasts with the situation in the often assumed case of infinite ionospheric (or ground) conductivity, in which case the phases are constant, apart from discontinuous shifts of π radians where the amplitude goes to zero.

In addition, the wave oscillates with nominal nodes or antinodes of electric field at the ionospheres, depending on how the wave admittance there compares with the ionospheric conductance. Also, when the wave admittance and ionospheric conductance come nearest to matching, maximum damping occurs.

ALLAN and KNOX (1979a, b) brought together various ideas of previous authors in a self-consistent way. In their model, hydromagnetic resonances occur in a cavity of plasma embedded in a dipole geomagnetic field, having a composite boundary comprising the anisotropically conducting ionosphere, the insulating atmosphere, and an infinitely conducting ground. In the composite boundary the ionospheric conductance dominates.

As mentioned already, earlier work suggested that an anisotropically conducting ionosphere would couple a toroidal to an evanescent poloidal mode; but the characteristics of the toroidal mode, which in this case is the dominant mode throughout most of the magnetosphere, would be controlled by the Pedersen conductance alone. On this basis Allan and Knox believed that the particular set of analytic solutions of the hydromagnetic equations they obtained, assuming zero Hall conductance, would give a reasonably valid description of one set of hydromagnetic standing-wave modes in the real magnetosphere: viz. the toroidal-like modes.

These analytic solutions confirm, and extend to the case of asymmetric conductance at geomagnetically conjugate points, the results of Newton *et al.* on phase variation, node pattern and damping.

The assumed negligible effect of the Hall conductance on the toroidal mode is supported by the results of CRAVEN and LAWRIE (1979). These authors derived and numerically solved a dispersion equation which included finite Hall as well as Pedersen conductance, but in a simpler geometry than dipole. They found the frequency and damping of a standing Alfvén wave to be changed by only a few percent when the finite, realistically valued Hall conductance was reduced to zero.

3. Current View

The current view of hydromagnetic resonances in the magnetosphere, which has developed out of the work just outlined, will now be summarised.

There are two standing wave modes involving distortion of the geomagnetic field, imagined frozen to the magnetospheric plasma. In one mode (the poloidal) the distortion of the field involves radial motion, while the other (the toroidal) involves azimuthal motion.

In general the two modes are coupled by field compression, but in the special case of axisymmetry (Fig. 1) and an isotropically conducting ionosphere they are independent. In this special case energy in the poloidal mode flows both along the geomagnetic field and radially across it, but in the toroidal mode (now with no field compression) it flows only along the geomagnetic field.

3.1 Details of toroidal resonance

The picture of a toroidal resonance we now have is as shown in Fig. 2, for the particular case of a near-node of electric field in the ionosphere, and when E_y is maximum at the resonant geomagnetic field line. There is a current loop, composed of geomagnetic field-aligned currents closed in the magnetosphere by displacement current, and in the ionosphere by Pedersen current. These currents occur as extended sheets and confine the azimuthal magnetic induction within the loop, so that it is screened from the ground. Associated with this current loop is an ionospheric Hall current which gives rise to a new magnetic induction in the northsouth direction, detectable at the ground. Thus there is an *apparent* 90° rotation of the wave magnetic component in going from the magnetosphere to the ground: left-handed in the northern and right-handed in the southern hemisphere. However if the east-west scale is only moderately large, some east-west magnetic induction should reach ground level, making the rotation somewhat different from 90°.

At the finitely-conducting ionosphere, on the resonant geomagnetic field line, the electric field and magnetic induction are in phase. As we move out along the line toward the equatorial plane, the phase of the electric field steadily changes by up to $\pi/2$ radians,

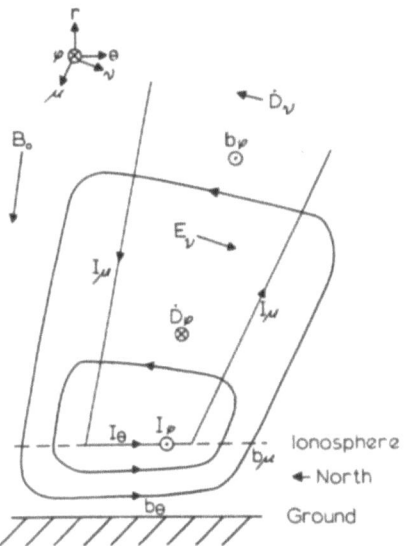

Fig. 2. Meridian cross-section of the axisymmetric resonant toroidal and coupled evanescent poloidal modes near the northern hemisphere ionosphere, at the instant when E_ν is maximum at the resonant geomagnetic field line.

but the phase of the magnetic induction stays nearly constant.

For the case of an antinode of electric field in the ionosphere the pattern is rather different. In particular the phase of the *magnetic induction* now changes along the resonant field line, while the phase of the *electric field* stays nearly constant.

Resonances involving a half-wave fundamental, with near-nodes of electric field in conjugate ionospheres, are expected to be common. However resonances with anti-nodes of electric field in the ionosphere are also expected where and when the ionosphere Pedersen conductance is sufficiently low. Such low conductance is not particularly common. Thus half-wave resonances with an antinode of electric field in the ionosphere of both hemispheres are less likely to occur than quarter-wave resonances with an antinode in the ionosphere of only one hemisphere and a near-node in the other. For a given dipole shell and plasma density, the quarter-wave fundamental resonance period is twice that of the half-wave.

Numerical damping values for toroidal and guided poloidal modes, when the plasma density is proportional to r^{-3}, are given by NEWTON *et al.* (1978). The following is an expression for toroidal mode damping in a plasma with density proportional to r^{-6} (ALLAN and KNOX, 1979a):

$$\gamma_m = \frac{\omega_0}{m\pi} \ln \left| \frac{(a_P^N - \zeta_I)(a_P^S - \zeta_I)}{(a_P^N + \zeta_I)(a_P^S + \zeta_I)} \right| .$$

3.2 *Coupling between modes*

Two types of coupling between modes exist. For non-axisymmetric modes one type occurs throughout the magnetosphere and is described by the following coupled equations (RADOSKI, 1967):

$$\left[\frac{\partial}{\partial \mu} \left(\frac{1}{h_\nu^2} \frac{\partial}{\partial \mu} \right) + \frac{\partial^2}{\partial \varphi^2} + k^2 h_\varphi^2 \right] (h_\nu E_\nu) = \frac{\partial^2}{\partial \nu \partial \varphi} (h_\varphi E_\varphi) ,$$

$$\left[\frac{\partial}{\partial\mu}\left(\frac{1}{h_\varphi^2}\frac{\partial}{\partial\mu}\right)+\frac{\partial^2}{\partial\nu^2}+k^2h_\nu^2\right](h_\varphi E_\varphi)=\frac{\partial^2}{\partial\nu\partial\varphi}(h_\nu E_\nu)\ .$$

It can be seen that in the axisymmetric case ($\partial/\partial\varphi\equiv0$) the equations decouple giving independent toroidal (E_ν alone) and poloidal (E_φ alone) modes.

The other type of coupling occurs in the ionosphere. Even for axisymmetry, if the ionospheric conductance is anisotropic, the toroidal and poloidal modes are coupled via the Hall conductance. The wave boundary conditions, including this coupling (for geomagnetic dip angle $\sim90°$), are as follows (ALLAN and KNOX, 1979a):

$$[a_P E_\nu^0+a_H E_\varphi^0=\pm\frac{1}{\mu_0}b_\varphi^0]^{S,N}\ ,$$

$$[a_P E_\varphi^0-a_H E_\nu^0=\mp\frac{1}{\mu_0}(b_\nu^0-b_\nu^1)]^{S,N}\ .$$

The upper sign goes with superscript S, and the lower with N.

3.3   Finite width toroidal standing wave

One set of axisymmetric modes presently of interest comprises a standing toroidal wave in the magnetosphere, coupled by the ionosphere to a poloidal wave, which is evanescent in the direction away from the ionosphere into the magnetosphere. Such an evanescent poloidal component is consistent with its associated thin toroidal component, and would be expected to transfer energy latitudinally across the geomagnetic field, spreading out the toroidal component from an infinitesimally thin to a finite (though still relatively thin) oscillating shell. This coupled mode meets Hruška's earlier mentioned objection to the infinitesimally thin shell.

From general consideration of forced oscillation of potentially resonant structures, the finite thickness shell is expected to oscillate at a single frequency, with maximum amplitude where this frequency coincides with the local resonant frequency. The width of the shell at the equatorial plane is expected to be of order $\gamma R_E/(\partial\omega_0/\partial L)$ (NEWTON et al., 1978).

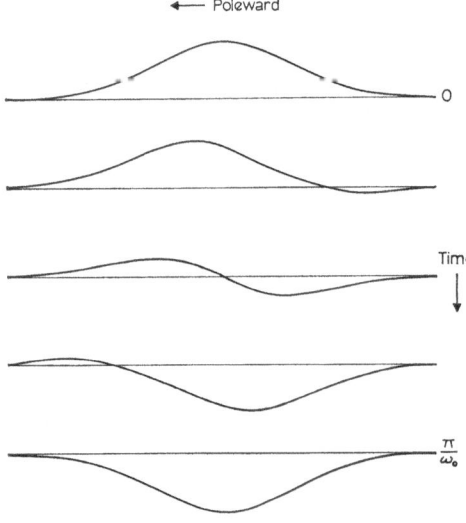

Fig. 3. Amplitude of E_ν versus geomagnetic latitude in the ionosphere, plotted within a resonant shell for five instants in a sequence extending over half a wave cycle.

It is also expected that the phase will vary continuously with latitude by up to π radians from one side of the resonating shell to the other.

The latitudinal phase variation in the ionosphere gives rise to a peculiar time sequence in the electric field standing wave pattern, as shown in Fig. 3 over a half-cycle interval. Momentarily the electric field is everywhere in one direction only, but then changes sign at the equatorward edge of the shell (where the amplitude is infinitesimal). The leading edge of the region of reversed electric field sweeps poleward across the shell until half a cycle later the whole field is momentarily reversed. At the equatorward edge a further reversal of electric field occurs, which, half a cycle later again, results in the original pattern being restored.

This continual poleward movement of the electric field pattern within the oscillating shell was first explained in terms of phase variation across the shell in the ionosphere by WALKER *et al.* (1979), in interpreting auroral radar studies of geomagnetic pulsations.

3.4   Magnification of electric field due to dipole geometry

The observation by auroral radar of large pulsation-associated electric fields in the ionosphere (UNWIN and KNOX, 1971) previously appeared to conflict with other considerations suggesting electric field nodes there. However this can now be explained as due to a magnifying effect of the dipole geometry on the electric field, due to the geomagnetic field lines converging as the ionosphere is approached.

By way of illustration the particular case is taken of a half-wave resonance along $L=4$ with $a_P^{S,N}=10\,S$: this has near-nodes of electric field in the ionospheres. If the geometrical factor (h_ν^{-1}) is removed from the expression for the electric field (ALLAN and KNOX, 1979a), the reasonable looking dashed line in Fig. 4 is obtained. However if the

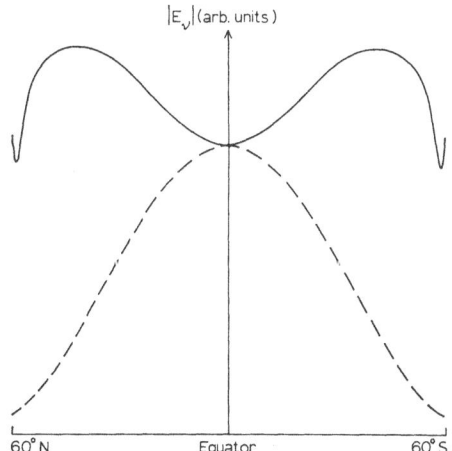

Fig. 4. Variation of electric field amplitude along the geomagnetic field in a half-wave resonance ($L=4$, $a_P^{S,N}=10\,S$) with near nodes of electric field at the ionospheres (solid line); and what it would be without the dipole geometric factor (dashed line).

factor is left in, the solid line is obtained, which is how the electric field should actually vary along the geomagnetic field. Note that at the ionosphere the amplitude of the actual electric field is comparable with that at the equatorial plane, thus resolving the apparent conflict.

4. Comparison with Observation

By and large, a wide variety of observations supports the picture of damping and coupling just given. On the other hand these observations also show a significant, but usually small, b_φ below the ionosphere. If the resonant mode were axisymmetric, as assumed in Section 3, no b_φ below the ionosphere would be expected. Thus observation implies a non-zero azimuthal wave-number to allow some leakage of b_φ from above the ionosphere to below.

Supporting observations (of which there are many examples in the References) include the relation between the electric field in the ionosphere and the magnetic induction at ground level, the relation between the polarisation of the magnetic component of the wave in the magnetosphere and the component at ground level, and the variation of polarisation and phase with latitude. These observations have been principally satellite and ground-based magnetometer readings, satellite measurements of plasma velocity, ground-based auroral radar measurements, and measurement of periodically modulated VLF waves propagating through the magnetosphere. Comparison of wave periods with measured or expected plasma mass densities at the appropriate dipole shell agree moderately well with theory, if quarter-wave resonances are included; but in the plasmatrough, even for quarter-wave resonances, periods are often higher than would be expected for an assumed hydrogen plasma. Explanations offered have been: (a) inflation of the geomagnetic field lines away from the dipole shape (WALKER et al., 1979), or (b) increased plasma mass density. The latter may be because of detached pieces of the plasmasphere drifting in the trough (WARNER and ORR, 1979), or the recently-observed presence of appreciable concentrations of ions heavier than hydrogen (e.g. SINGER et al., 1979).

5. Future Work

Finally, the authors would like to make some suggestions for future work:

1) Further development of the theory of ionospheric coupling between axisymmetric resonant toroidal and evanescent poloidal waves, and between axisymmetric resonant poloidal waves and toroidal waves. Generalisation to the non-axisymmetric case with magnetospheric coupling is also required. The theory should include partial reflection of the poloidal wave at the plasmapause.

2) Further measurement of the difference in polarisation of the magnetic component of the wave below and just above the ionosphere. This could give information on the east-west extent of the wave, or the azimuthal wave number.

3) Accurate measurement of phase variation along the geomagnetic field, particularly the phase difference between the wave electric and magnetic components. Among other things they may allow us to distinguish unambiguously quarter- from half-wave resonances.

4) Further accurate and reasonably spaced measurements of wave profiles (particularly of phase) in both the north-south and east-west directions. Auroral radar is very useful in this connection.

5) Comparison of observed resonance widths with damping predicted from ionospheric Pedersen conductance.

6) In conclusion, ideally all pulsation measurements should be accompanied by

other measurements allowing deduction of ionospheric conductance and magnetospheric mass density; and, if possible, latitudinal or radial variation of such. Also, measurement of ground distortion of magnetic field is necessary to properly interpret sub-ionospheric measurements. We realise this is counsel of perfection in a very imperfect world, but is worth keeping in mind when designing experiments.

REFERENCES A

Works cited in text.

ALLAN, W. and F. B. KNOX, A dipole field model for axisymmetric Alfvén waves with finite ionosphere conductivities, *Planet. Space Sci.*, **27**, 79–85, 1979a (see also "Erratum", *ibid.*, p. 1045).

ALLAN, W. and F. B. KNOX, The effect of finite ionosphere conductivities on axisymmetric toroidal Alfvén wave resonances, *Planet. Space Sci.*, **27**, 939–950, 1979b.

CRAVEN, A. H. and J. A. LAWRIE, The effect of the lower ionosphere on hydromagnetic modes in the plasmasphere, *Planet. Space Sci.*, **27**, 211–213, 1979.

DUNGEY, J. W., Electrodynamics of the outer atmosphere, Scientific Report No. 69, Pennsylvania State University, U.S.A., 1954.

DUNGEY, J. W., *Cosmic Electrodynamics*, pp. 169–172, Cambridge University Press, 1958.

DUNGEY, J. W., Hydromagnetic waves and the ionosphere, Proc. Int. Conference on the Ionosphere, pp. 230–232, Inst. Physics, London, 1963.

HRUŠKA, A., The magnetodynamic toroidal waves, *Planet. Space Sci.*, **16**, 1305–1309, 1968.

HUGHES, W. J., The effect of the atmosphere and ionosphere on long period magnetospheric micropulsations, *Planet. Space Sci.*, **22**, 1157–1172, 1974.

HUGHES, W. J. and D. J. SOUTHWOOD, The screening of micropulsation signals by the atmosphere and ionosphere, *J. Geophys. Res.*, **81**, 3234–3240, 1976a.

HUGHES, W. J. and D. J. SOUTHWOOD, An illustration of modification of geomagnetic pulsation structure by the ionosphere, *J. Geophys. Res.*, **81**, 3241–3247, 1976b.

INOUE, Y., Wave polarizations of geomagnetic pulsations observed in high latitudes on the earth's surface, *J. Geophys. Res.*, **78**, 2959–2976, 1973.

MAL'TSEV, Yu.P. and S. V. LEONT'YEV, Ionospheric conductivity and attenuation of Pi 2 pulsations, *Geomagn. Aeron.*, **17**, 95–96, 1977 (English transl.).

NEWTON, R. S., D. J. SOUTHWOOD, and W. J. HUGHES, Damping of geomagnetic pulsations by the ionosphere, *Planet. Space Sci.*, **26**, 201–209, 1978.

NISHIDA, A., Ionospheric screening effect and storm sudden commencement, *J. Geophys. Res.*, **69**, 1861–1874, 1964.

RADOSKI, H. R., A note on oscillating field lines, *J. Geophys. Res.*, **72**, 418–419, 1967.

SINGER, H. J., C. T. RUSSELL, M. G. KIVELSON, T. A. FRITZ, and W. LENNARTSON, Satellite observations of the spatial extent and structure of Pc 3, 4, 5 pulsations near the magnetospheric equator, *Geophys. Res. Lett.*, **6**, 889–892, 1979.

UNWIN, R. S. and F. B. KNOX, Radio aurora and electric fields, *Radio Sci.*, **6**, 1061–1077, 1971.

VAN'YAN, L. L. and L. A. ABRAMOV, Propagation of hydromagnetic waves, guided by the magnetic field, through the lower layers of the ionosphere, *Geomagn. Aeron.*, **9**, 121–122, 1969a (English transl.).

VAN'YAN, L. L. and L. A. ABRAMOV, Propagation of low-frequency hydromagnetic waves, guided by the geomagnetic field, through a gyrotropic ionosphere, *Geomagn. Aeron.*, **9**, 620–621, 1969b (English transl.).

VAN'YAN, L. L. and A. A. KOZHEVNIKOV, Effect of the geomagnetic field inclination on the reflection from the ionosphere of guided hydromagnetic waves, *Geomagn. Aeron.*, **14**, 256–261, 1974 (English transl.).

VAN'YAN, L. L., M. B. GOKHBERG, and L. A. ABRAMOV, Influence of the lower ionosphere on the propagation of hydromagnetic waves directed by the geomagnetic field, *Geomagn. Aeron.*, **10**, 565–568, 1970 (English transl.).

Walker, A. D. M., R. A. Greenwald, W. F. Stuart, and C. A. Green, STARE auroral radar observations of Pc 5 geomagnetic pulsations, *J. Geophys. Res.*, **84**, 3373–3388, 1979.

Warner, M. R. and D. Orr, Time of flight calculations for high latitude geomagnetic pulsations, *Planet. Space Sci.*, **27**, 679–689, 1979.

REFERENCES   B

A selection for further reading.

Allan, W. and F. B. Knox, Interpretation of an ATS 6 Alfvén wave using solutions with finite ionosphere conductivity, *Geophys. Res. Lett.*, **5**, 849–852, 1978.

Andrews, M. K., Magnetic pulsation behaviour in the magnetosphere inferred from whistler mode signals, *Planet. Space Sci.*, **25**, 957–966, 1977.

Andrews, M. K., L. J. Lanzerotti, and C. G. Maclennan, Rotation of hydromagnetic waves between magnetosphere and ground, *J. Geophys. Res.*, **84**, 7267–7270, 1979.

Arthur, C. W., R. L. McPherron, and W. J. Hughes, A statistical study of Pc 3 magnetic pulsations at synchronous orbit, ATS 6, *J. Geophys. Res.*, **82**, 1149–1157, 1977.

Arykov, A. A. and Yu.P. Mal'tsev, Artificial oscillations of geomagnetic field line, *Planet. Space Sci.*, **27**, 463–471, 1979.

Barfield, J. N., N. M. Bondarenko, A. M. Buloshnikov, M. B. Gokhberg, A. L. Kalisher, R. L. McPherron, and V. A. Troitskaya, Synchronous observations of long-period geomagnetic pulsations on the ATS 6 satellite and at the surface of the earth, *Geomagn. Aeron.*, **17**, 596–599, 1977 (English transl.).

Beamish, D., H. W. Hanson, and D. C. Webb, Complex demodulation applied to Pi 2 geomagnetic pulsations, *Geophys. J. R. Astr. Soc.*, **58**, 471–493, 1979.

Craven, A. H. and J. A. Lawrie, A simple mechanical analogy to illustrate a latitude phase effect in mid-latitude micropulsations, *Planet. Space Sci.*, **24**, 612–613, 1976.

Cummings, W. D., S. E. DeForest, and R. L. McPherron, Measurements of the Poynting vector of standing hydromagnetic waves at geosynchronous orbit, *J. Geophys. Res.*, **83**, 697–706, 1978.

Davydov, V. M., Effect of the time-dependence of the lower ionosphere on the parameters of the magnetospheric resonator, *Geomagn. Aeron.*, **16**, 156–159, 1976 (English transl.).

Davydov, V. M. and L. F. Snegurova, Influence of the lower layers of the ionosphere and of the earth on the field of three-dimensional Alfvén waves, *Geomagn. Aeron.*, **10**, 710–713, 1970 (English transl.).

Fukunishi, H. and L. J. Lanzerotti, ULF pulsation evidence of the plasmapause 2. Polarization studies of Pc 3 and Pc 4 pulsations near $L=4$ and at a latitude network in the conjugate region, *J. Geophys. Res.*, **79**, 4632–4647, 1974.

Fukunishi, H., L. J. Lanzerotti, and C. G. Maclennan., Three-dimensional polarization characteristics of magnetic variations in the Pc 5 frequency range at conjugate areas near $L=4$, *J. Geophys. Res.*, **80**, 3973–3984, 1975.

Green, C. A., Meridional characteristics of a Pc 4 micropulsation event in the plasmasphere, *Planet. Space Sci.*, **26**, 955–967, 1978.

Green, C. A. and R. A. Hamilton, Polarization characteristics and phase differences of Pi 2 pulsations at conjugate stations, *J. Atmos. Terr. Phys.*, **40**, 1223–1228, 1978.

Hanson, H. W., D. C. Webb, and D. Beamish, A high resolution study of continuous pulsations in the European sector, *Planet. Space Sci.*, **27**, 1371–1382, 1979.

Hasegawa, A. and L. J. Lanzerotti, On the orientation of hydromagnetic waves in the magnetosphere, *Rev. Geophys. Space Phys.*, **16**, 263–266, 1978.

Hughes, W. J., R. L. McPherron, and J. N. Barfield, Geomagnetic pulsations observed simultaneously on three geostationary satellites, *J. Geophys. Res.*, **83**, 1109–1116, 1978.

Kokubun, S., R. L. McPherron, and C. T. Russell, Ogo 5 observations of Pc 5 waves: ground-magnetosphere correlations, *J. Geophys. Res.*, **81**, 5141–5149, 1976.

Lanzerotti, L. J., C. G. Maclennan, and H. Fukunishi, ULF geomagnetic power near $L=4$, 5: Cross-power spectral studies of geomagnetic variations 2–27 mHz in conjugate areas, *J. Geophys. Res.*, **81**,

3299–3315, 1976.

MACLENNAN, C. G., L. J. LANZEROTTI, A. HASEGAWA, E. A. BERING III, J. R. BENBROOK, W. R. SHELDON, T. J. ROSENBERG, and D. L. MATTHEWS, On the relationship of ~3 mHz (Pc 5) electric, magnetic and particle variations, *Geophys. Res. Lett.*, **5**, 403–406, 1978.

MALLINCKRODT, A. J. and C. W. CARLSON, Relations between transverse electric fields and field-aligned currents, *J. Geophys. Res.*, **83**, 1426–1432, 1978.

MAL'TSEV, Yu.P., Boundary condition for Alfvén waves at the ionosphere, *Geomagn. Aeron.*, **17**, 677–679, 1977 (English transl.).

MAL'TSEV, Yu.P., S. V. LEONT'YEV, and V. B. LYATSKY, Generation and natural frequencies of Pi 2 pulsations, *Geomagn. Aeron.*, **14**, 101–107, 1974a (English transl.).

MAL'TSEV, Yu.P., S. V. LEONT'YEV, and V. B. LYATSKY, Pi-2 pulsations as a result of evolution of an Alfvén impulse originating in the ionosphere during a brightening of aurora, *Planet. Space Sci.*, **22**, 1519–1533, 1974b.

MAL'TSEV, Yu. P., V. B. LYATSKY, and A. M. LYATSKAYA, Currents over the auroral arc, *Planet. Space Sci.*, **25**, 53–57, 1977.

McDIARMID, D. R., Comment on "A generation mechanism for Pc 5 micropulsations in the morning sector" by Gordon Rostoker and Hing-Lan Lam, *Planet. Space Sci.*, **27**, 1123–1125, 1979.

MIER-JEDRZEJOWICZ, W. A. C. and D. J. SOUTHWOOD, The east-west structure of mid-latitude geomagnetic pulsations in the 8–25 mHz band, *Planet. Space Sci.*, **27**, 617–630, 1979.

RIETVELD, M. T., R. L. DOWDEN, and L. E. S. AMON, Micropulsations observed by whistler-mode transmissions, *Nature*, **276**, 165–167, 1978.

ROSTOKER, G. and H.-L. LAM, A generation mechanism for Pc 5 micropulsations in the morning sector, *Planet. Space Sci.*, **26**, 493–505, 1978.

SAMSON, J. C., J. A. JACOBS, and G. ROSTOKER, Latitude-dependent characteristics of long-period geomagnetic micropulsations, *J. Geophys. Res.*, **76**, 3675–3683, 1971.

SOUTHWOOD, D. J. and W. J. HUGHES, Source induced vertical components in geomagnetic pulsation signals, *Planet. Space Sci.*, **26**, 715–720, 1978.

STUART, W. F., P. M. BRETT, and T. J. HARRIS, Mid-latitude secondary resonance in Pi 2's, *J. Atmos. Terr. Phys.*, **41**, 65–75, 1979.

VAN'YAN, L. L. and L. A. ABRAMOV, Influence of the inclination of geomagnetic lines of force on the propagation of guided hydromagnetic waves in the lower ionosphere, *Geomagn. Aeron.*, **9**, 733–735, 1969 (English transl.).

VAN'YAN, L. L. and M. B. GOKHBERG, Excitation of magnetospheric resonators, *Geomagn. Aeron.*, **12**, 435–440, 1972 (English transl.).

The Rotation of Hydromagnetic Waves by the Ionosphere

M. K. Andrews,* L. J. Lanzerotti,** and C. G. Maclennan**

*Physics and Engineering Laboratory, DSIR, Lower Hutt, New Zealand
**Bell Laboratories, Murray Hill, New Jersey, U.S.A.

(Received June 28, 1980)

We show that the Doppler shift on whistler mode signals from VLF transmitters may be used to infer the magnetic field of an hydromagnetic wave above the ionosphere in the magnetosphere. This inferred field can then be compared with that measured on the ground. We examine one event in detail and show the data imply that a $\pi/2$ rotation of the orientation of the wave ellipse has occurred between the magnetosphere and the ground.

1. Introduction

In this paper we summarize, from the work of Andrews (1977) and Andrews et al. (1979), how VLF whistler mode signals, propagating in the magnetosphere, can be used with ground-based recordings of hydromagnetic waves (magnetic pulsations) to demonstrate the existence of a rotation of the wave ellipse orientation between the magnetosphere and ground, as predicted earlier by several authors (e.g., Inoue, 1973; Hughes 1974).

Whistler mode (W/M) signals from high-power VLF transmitters in the northern hemisphere have been received in New Zealand and Antarctica for a number of years. The transmitted frequency is precisely known; the received signal is not very coherent and is usually Doppler shifted, implying the phase path in the whistler duct (in the magnetosphere) between receiver and transmitter is rarely constant. This means that either refractive index changes are occurring within the duct, or that the duct length is changing, usually as a result of cross-L drifts.

In addition to cross-L drift, it has been found that the Doppler shift gives a sensitive

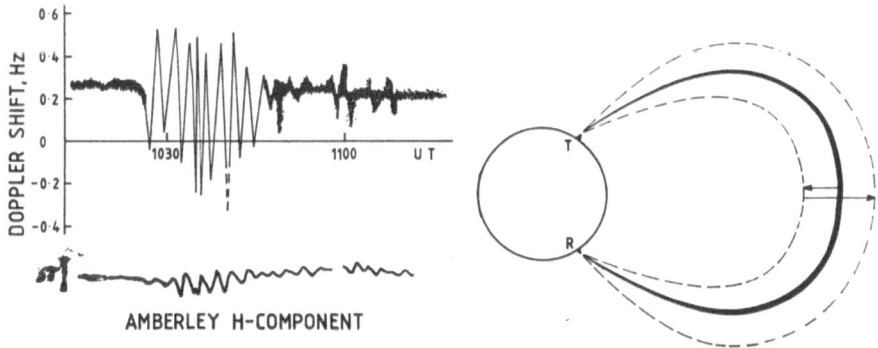

Fig. 1. Oscillation of Doppler shift of fixed frequency VLF whistler mode signal produced by a magnetic pulsation observed on October 21, 1966.

indication of the radial motion of a duct under the influence of an hydromagnetic wave (ANDREWS, 1977). Figure 1 shows an example of the phenomenon. Whistler-mode signals from NLK, Seattle (18.6 kHz) were being received at Wellington with a Doppler shift of 0.2 Hz on October 21, 1966. At 1027 UT, the W/M signal began executing a number of oscillations. Magnetograms from Amberley, near Christchurch, showed the simultaneous onset of a magnetic pulsation (Pc 4 type). The effect is explained in terms of standing poloidal oscillations of the W/M duct, oscillations which cause the VLF phase path to increase and decrease in time with the pulsation (ANDREWS, 1977). The existence of such effects on the VLF W/M signals is strong evidence in support of a standing wave interpretation of magnetic pulsations.

Toroidal (i.e., azimuthal) oscillations of a duct will scarcely affect the VLF phase path (ANDREWS, 1977; ANDREWS et al., 1979). Therefore, such sets of observations constitute a ground-based method of identifying the phase of the standing hydromagnetic wave in the magnetosphere, which can then be compared with ground magnetic data.

Figure 2 outlines the method of analysis. If the mode of oscillation is half wave with nodes in the ionosphere, the field line (or duct) in its earthward position can be represented by the addition of a transverse standing wave field to the dipole line, as illustrated in Fig. 2.

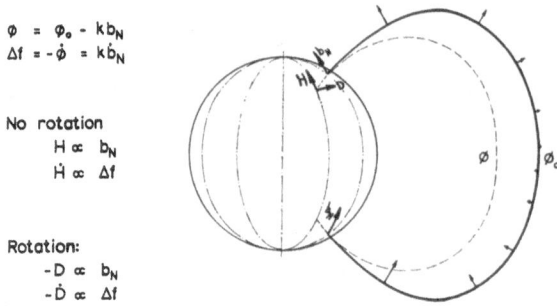

Fig. 2.   Relation of VLF phase path to pulsation signals on the ground.

The radial wave field just above the ionosphere is northward and is denoted by b_N. The result is a shorter VLF duct, with a phase path length ϕ (which is loosely related to the physical path length) less than the phase path ϕ_0 of the unperturbed, dipolar magnetic field line. Thus, if k is a positive constant (whose value will depend on the duct latitude)

$$\phi = \phi_0 - kb_N$$

$$\Delta f = -\dot{\phi} = kb_N .$$

If the ionosphere does not rotate the orientation of the wave ellipse, the pulsation signal b_N will be seen on the ground as a northerly component; i.e.,

$$H \propto b_N$$

so that

$$\dot{H} \propto \Delta f . \tag{1}$$

On the other hand, if the ionosphere does produce the rotation predicted by theory, b_{N} will be observed on the ground as a westerly component in the northern hemisphere, i.e.,

$$-D \propto b_{\mathrm{N}}$$

$$-\dot{D} \propto \Delta f. \tag{2}$$

From (1) and (2), the existence, or otherwise, of a signal rotation may be tested by comparing the Doppler shift Δf with the time differentials of the measured ground magnetic components.

A preliminary attempt to do this was made by ANDREWS (1977), but accuracies were limited by the magnetic data available. Subsequently, W/M recordings made at Siple, Antarctica, have shown the existence of many pulsation signals on the dayside during southern winter, and accurate vector magnetometer data is available from the Bell Labs chain in Antarctica and North America. Here we show one long-lived VLF W/M event and determine that the ground magnetic signal is rotated compared to that in the magnetosphere. The work is reported in detail elsewhere (ANDREWS *et al.*, 1979).

2. Data

The pulsation event illustrated here occurred near local noon on 6 July 1975. The mean period was 140 s, and magnetic activity was low. Figure 3 shows polarization data for the event from the magnetometer chain. At the time of interest (16–17 UT) there is a polarization reversal in the hydromagnetic waves at about $L=3.5$. This suggests a spatial resonance occurred near $L=3.5$, and is supported by a pulsation amplitude peak at this latitude. Plasmaspheric electron densities would be consistent with the oscillation mode being the fundamental. The VLF W/M duct was near $L=3$; i.e., on the low latitude

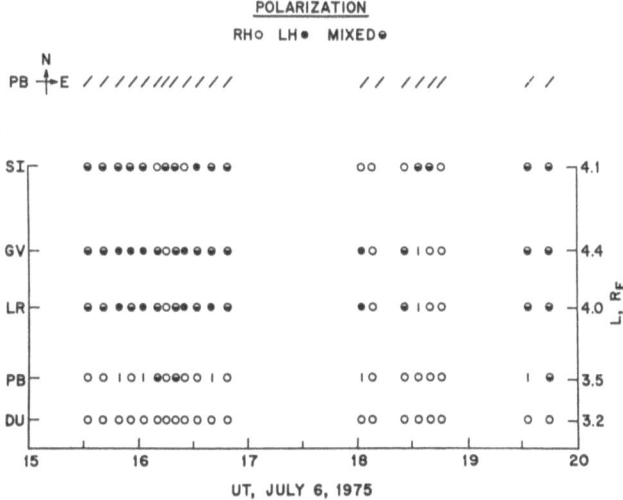

Fig. 3. Polarization data from the magnetometer chain. DU=Durham, PB=Pittsburg, LR=Lac Rebours, GV=Girardville, SI=Siple. The data at the top, for PB, indicate the orientation, in the horizontal plane, of the wave ellipse.

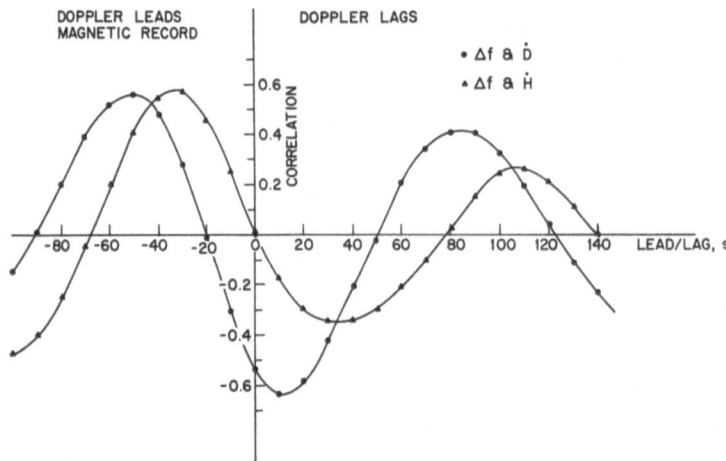

Fig. 4.   Cross correlation of Siple whistler mode signals and magnetic data from Durham, hour 16, July 6, 1975.

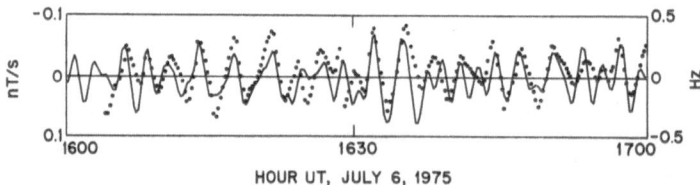

Fig. 5.   Superposition of $-\dot{D}$ and VLF Doppler shift to show agreement over a one hour interval.

side of the resonance region.

For the hour 16–17 UT, when the VLF data were unbroken, a cross correlation was done between Δf and $\dot{H}$ and $\dot{D}$ from Durham ($L=3.2$), the station nearest the foot of the VLF duct.   Figure 4 shows the result.   The periodic nature of the correlation reflects the quasi-sinusoidal character of the pulsation.   Near zero time lag, Δf and $\dot{H}$ show zero correlation, but Δf is highly anticorrelated with D; i.e.,

$$\Delta f \propto -\dot{D}$$

which is the relation indicating that the signal in the magnetosphere is observed on the ground with a 90° rotation.   (Maximum anticorrelation is actually reached at $+11$ s, but this is within the ± 15 s timing accuracy of the VLF records.)

In Fig. 5, $-\dot{D}$ and Δf are superimposed to show the degree of agreement, which is considered to be good.

Figure 6 shows a frame of data occurring one hour later, when W/M data were particularly well defined, and illustrates, on a non-statistical basis, the better agreement between $-\dot{D}$ and Δf than between $-\dot{H}$ and Δf.   Note in particular that the VLF trace near the eighth minute is matched better by the $-\dot{D}$ trace than by the $\dot{H}$ trace.   The latter

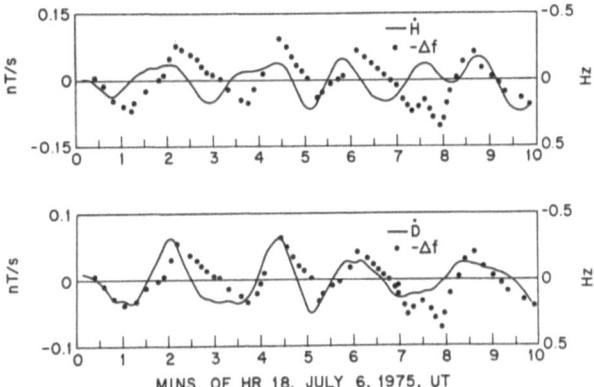

Fig. 6. Comparison of a well-defined W/M record with Durham magnetometer data. The agreement of Δf and $-\dot{D}$ implies an ionospheric rotation. (See note added in proof).

shows an extra oscillation at this time not present on $\dot{D}$.

We draw two conclusions from this study. First, the existence of pulsation effects in the Doppler shift of fixed-frequency VLF W/M records is strong evidence in support of a standing wave interpretation of magnetic pulsation signals. Second, the VLF signal can be used to give information on the phase of the hydromagnetic wave signal in the magnetosphere. Our data show that the northward signal in the magnetosphere matches the westward signal on the ground; i.e., a rotation in wave orientation has occurred, as predicted by theory. We note that a similar result on wave orientation was achieved using the STARE radar in Scandinavia to measure pulsation electric fields in the ionosphere (WALKER et al., 1979).

REFERENCES

ANDREWS, M. K., Magnetic pulsation behaviour in the magnetosphere inferred from whistler mode signals, *Planet. Space Sci.*, **25**, 957–966, 1977.

ANDREWS, M. K., L. J. LANZEROTTI, and C. G. MACLENNAN, Rotation of hydromagnetic waves between the magnetosphere and the ground, *J. Geophys. Res.*, **84**, 7267–7270, 1979.

HUGHES, W. J., The effect of the atmosphere and ionosphere on long period magnetospheric micropulsations, *Planet. Space Sci.*, **22**, 1157–1172, 1974.

INOUE, Y., Wave polarizations of geomagnetic pulsations observed in high latitudes on the earth's surface, *J. Geophys. Res.*, **78**, 2959–2976, 1973.

WALKER, A. D. M., R. A. GREENWALD, W. F. STUART, and C. A. GREEN, Stare auroral radar observations of Pc 5 geomagnetic pulsations, *J. Geophys. Res.*, **84**, 3373–3388, 1979.

Note added in proof

The upper diagram in Fig. 6 should have $-\dot{H}$ superimposed on $-\Delta f$ for comparison with Eq. (1). If this is done the disagreement between $\dot{H}$ and Δf is very obvious.

AEPS Vol. 1

Hard cover edition to Journal of Geomagnetism and Geoelectricity Vol. 29, No. 4, 1977

Proceedings of AGU 1976 Fall Annual Meeting, December 1976, San Francisco

ORIGIN OF THERMOREMANENT MAGNETIZATION

Edited by David J. DUNLOP

Contents TRM and Its Variation with Grain Size: A Review (R. DAY)/Single Domain Oxide Particles as a Source of Thermoremanent Magnetization (M.E. EVANS)/Domain Structure of Titanomagnetities and Its Variation with Temperature (H.C. SOFFEL)/The Demagnetization Field of Multidomain Grains (R.T. MERRILL)/The Hunting of the 'Psark' (D.J. DUNLOP)/On the Origin of Stable Remanence in Pseudo-Single Domain Grains (S.K. BANERJEE)/The Preparation, Characterization and Magnetic Properties of Synthetic Analogues of Some Carriers of the Palaeomagnetic Record (J.B. O'DONOVAN and W. O'REILLY)/Reduction of Hematite to Magnetite under Natural and Laboratory Conditions (P.N. SHIVE and J.F. DIEHL)/Characteristics of First Order Shock Induced Magnetic Transitions in Iron and Discrimination from TRM (P. WASILEWSKI)/The Thermoremanence Hypothesis and the Origin of Magnetization in Iron Meteorites (A. BRECHER and L. ALBRIGHT)/Thermal Overprinting of Natural Remanent Magnetization and K/Ar Ages in Metamorphic Rocks (K.L. BUCHAN, G.W. BERGER, M.O. MCWILLIAMS, D. YORK, and D.J. DUNLOP)/Does TRM Occur in Oceanic Layer 2 Basalts? (J.M. HALL)/The Effects of Alteration on the Natural Remanent Magnetization of Three Ophiolite Complexes: Possible Implications for the Oceanic Crust (S. LEVI and S.K. BANERJEE)

AEPS Vol. 2

Hard cover edition to Journal of Physics of the Earth Vol. 25, Supplement, 1977 (Not included in regular issues)

Proceedings of the U.S.-Japan Seminar on Theoretical and Experimental Investigations of Earthquake Precursors

EARTHQUAKE PRECURSORS

Edited by C. KISSLINGER and Z. SUZUKI

Contents Earthquake Prediction-Related Research at the Seismological Laboratory, California Institute of Technology, 1974–1976 (J.H. WHITCOMB)/Research on Earthquake Prediction and Related Areas at Columbia University (L.R. SYKES)/Seismic Activities and Crustal Movements the Yamasaki Fault and Surrounding Regions in the Southwest Japan (K. OIKE)/The New Madrid Seismic Zone as a Laboratory for Earthquake Prediction Research (B.J. MITCHELL, W. STAUDER, and C.C. CHENG)/Anomalous Crustal Activity in the Izu Peninsula, Central Honshu (K. TSUMURA)/Recent Seismometrical Works in Japan (S. SUYEHIRO, M. ICHIKAWA, and K. TSUMURA)/Quiet and Violence in Horizontal Movement of the Crust (T. HARADA)/Anomalous Seismic Activity and Earthquake Prediction (H. SEKIYA)/Seismic Activity in the Northeastern Japan Arc (A. TAKAGI, A. HASEGAWA, and N. UMINO)/Observations of Changes in Seismic Wave Velocity in South Kanto District, South of Tokyo, by the Explosion-Seismic Method (T. KAKIMI and I. HASEGAWA)/Some Precursors Prior to Recent Great Earthquakes along the Nankai Trough (H. SATO)/Possibility of Temporal Variations in Earth Tidal Strain Amplitudes Associated with Major Earthquakes (T. MIKUMO, M. KATO, H. DOI, Y. WADA, T. TANAKA, R. SHICHI, and A. YAMAMOTO)/Gravity Changes Associated with Seismic Activities (Y. HAGIWARA)/Geomagnetism in Relation to Tectonic Activities of the Earth's Crust in Japan (N. SUMITOMO) / Precursory and Coseismic Changes in Ground Resistivity (T. RIKITAKE and Y. YAMAZAKI) / Geochemistry as a Tool for Earthquake Prediction (H. WAKITA) / Recent Laboratory Studies of Earthquake Mechanics and Prediction (W.F. BRACE) / Dilatancy of Rocks under General Triaxial Stress States with Special Reference to Earthquake Precursors (K. MOGI) / Possibility of a Great Earthquake in the Tokai District, Central Japan (T. UTSU) / Depth Constraints on Dilatancy Induced Velocity Anomalies (K.W. WINKER and A. NUR) / Seismological Precursors to a Magnitude 5 Earthquake in the Central Aleutian Islands (E.R. ENGDAHL and C. KISSLINGER) / Estimation of Future Destructive Earthquakes from Active Faults on Land in Japan (T. MATSUDA) / Some Problems in the Prediction of the Nemuro-oki Earthquake (K. ABE) / Responses to Earthquake Prediction in Kawasaki City, Japan in 1974 (H. OHTA and K. ABE) / Socioeconomic and Political Consequences of Earthquake Prediction (J.E. HAAS and D.S. MILETI)

AEPS Vol. 3

Proceedings of the U.S.-Japan Seminar on Rare Gas Abundance and Isotopic Constraints on the Origin and Evolution of the Earth's Atmosphere

TERRESTRIAL RARE GASES

Edited by E.C. ALEXANDER, Jr. and M. OZIMA

Contents *EXPERIMENTAL STUDIES* A Mantle Helium Component in Circum-Pacific Volcanic Gases: Hakone, the Marianas, and Mt. Lassen (H. CRAIG, J.E. LUPTON, and Y. HORIBE) / Nitrogen to Argon Ratio in Volcanic Gases (S. MATSUO, M. SUZUKI, and Y. MIZUTANI) / Rare Gas Abundance Pattern of Fumarolic Gases in Japanese Volcanic Areas (O. MATSUBAYASHI, S. MATSUO, I. KANEOKA, and M. OZIMA) / A Review: Some Recent Advances in Isotope Geochemistry of Light Rare Gases (I.N. TOLSTIKHIN) / Abundances and Isotopic Compositions of Rare Gases in Granites and Thucholites (P.K. KURODA and R.D. SHERRILL) / Rare Gas Isotopic Compositions in Diamonds (N. TAKAOKA and M. OZIMA) / Rare Gases in Mantle-Derived Rocks and Minerals (I. KANEOKA, N. TAKAOKA, and K. AOKI) / A Comparison of Terrestrial and Meteoritic Noble Gases (O.K. MANUEL) / The Composition and History of the Martian Atmosphere (T. OWEN) *THEORETICAL STUDIES* Nuclear Components in the Atmosphere (T.J. BERNATOWICZ and F.A. PODOSEK) / Trapped Xenon and Cosmic-Ray Effects in Meteorites, in Lunar Sample, and in the Earth's Materials (K. SAKAMOTO) / Classification and Generation of Terrestrial Rare Gases (K. SAITO) / Earth-Atmosphere Evolution Model Based on Ar Isotopic Data (Y. HAMANO and M. OZIMA) / Terrestrial Potassium and Argon Abundances as Limits to Models of Atmospheric Evolution (D.E. FISHER) / On the Ambient Mantle $^4He/^{40}Ar$ Ratio and the Coherent Model of Degassing of the Earth (D.W. SCHWARTMAN) / Earth Degassing Models, and the Heterogenous vs. Homogeneous Mantle (R. HART and L. HOGAN) / Lead Isotope Constraints on the Early History of the Earth (R.D. RUSSELL) / Matter Accretion into the Solar System (S. HAYAKAWA)

AEPS Vol. 4

Hard cover edition to Journal of Geomagnetism and Geoelectricity Vol. 30, Nos. 3 and 4, 1978
Proceedings of IAGA/IAMAP Joint Assembly, August 1977, Seattle, Washington
AURORAL PROCESSES
Edited by C.T. RUSSELL

Contents *TIMING OF SUBSTORM EVENTS* Pi 2 Micropulsations as Indicators of Substorm Onsets and Intensifications (G. ROSTOKER and J.V. OLSON) / The Use of Ground Magnetograms to Time the Onset of Magnetospheric Substorms (R.L. MCPHERRON) / Substorm Onset in the Magnetotail (A. NISHIDA) *ELECTROMAGNETIC AND ELECTROSTATIC INSTABILITIES ON AURORAL FIELD LINES* A Review of Electrostatic Wave Measurements on Auroral Magnetic Field Lines (M.C. KELLEY) / Diffuse Auroral Precipitation (M. ASHOUR-ABDALLA and C.F. KENNEL) / Electromagnetic Plasma Wave Emissions from the Auroral Field Lines (D.A. GURNETT) / Theory of Electromagnetic Waves on Auroral Field Lines (J.E. MAGGS) *RAPID AURORAL FLUCTUATIONS AND ASSOCIATED PHENOMENON* Observations of Rapid Auroral Fluctuations (T. OGUTI) / Highlights in the Studies of the Relationship of Geomagnetic Field Changes to Auroral Luminosity (W.H. CAMPBELL) / Microburst Precipitation Phenomena (G.K. PARKS) *MECHANISMS FOR THE FORMATION OF AURORAL STRUCTURE* Observed Microstructure of Auroral Forms (T.N. DAVIS) / Birkeland Currents and Auroral Structure (H.R. ANDERSON) / Relationships between Particle Precipitation and Auroral Forms (J.L. BURCH and J.D. WINNINGHAM) / Photometric Investigation of Precipitating Particle Dynamics (S.B. MENDE) / Generation Mechanisms for Magnetic-Field-Aligned Electric Fields in the Magnetosphere (C.-G. FÄLTHAMMAR) / Review of Auroral Currents and Auroral Arcs (G. ATKINSON) / Acceleration Mechanisms for Auroral Electrons (D.W. SWIFT) / Subject Index

AEPS Vol. 5

Hard cover edition to Journal of Geomagnetism and Geoelectricity Vol. 30, No. 5, 1978

Proceedings of IAGA/IAMAP Joint Assembly, August 1977, Seattle, Washington

TECTONOMAGNETICS AND LOCAL GEOMAGNETIC FIELD VARIATIONS

Edited by M. FULLER, M.J.S. JOHNSTON, and T. YUKUTAKE

Contents Symposium on Tectonomagnetics and Small Scale Secular Variations Held at the IAGA/IAMAP Joint Assembly at Seattle on Tuesday, August 22nd, 1977 (V.A. SHAPIRO and M.J.S. JOHNSTON) / Tectonomagnetic Studies in Tajikstan (Yu. P. SKOVORODKIN, L.S. BEZUGLAYA, and T.V. GUSEVA) / An Attempt to Observe a Seismomagnetic Effect during the Gazly 17th May 1976 Earthquake (V.A. SHAPIRO and K.N. ABDULLABEKOV) / Secular Variation Anomalies and Aseismic Geodynamic in the Urals (V.A. SHAPIRO, A.L. ALEINKOV, A.A. NULMAN, V.A. PYANKOV, and A.V. ZUBKOV) / Geomagnetic Investigations in the Seismoactive Regions of Middle Asia (V.A. SHAPIRO, A.N. PUSHKOV, K.N. ABDULLABEKOV, E.B. BERDALIEV, and M.Yu. MUMINOV) / Local Magnetic Field Variations and Stress Changes Near a Slip Discontinuity on the San Andreas Fault (M.J.S. JOHNSTON) / Geomagnetic Secular Variation Anomalies in the GDR (W. MUNDT) / Noise Reduction Techniques for Use in Determining Local Geomagnetic Field Changes (R.H. WARE and P.L. BENDER) / Local Variations in Magnetic Field, Long-Term Changes in Creep Rate, and Local Earthquakes along the San Andreas Fault in Central California (B.E. SMITH, M.J.S. JOHNSTON, and R·O. BURFORD) / Geomagnetic Induction Study of the Seismically Active Fault along the Southwestern Coast of the Sea of Japan (J. MIYAKOSHI and A. SUZUKI) / Time Dependence of Magnetotelluric Fields in a Tectonically Active Region in Eastern Canada (R.D. KURTZ and E.R. NIBLETT) / Piezomagnetic Response with Depth, Related to Tectonomagnetism as an Earthquake Precursor (R.S. CARMICHAEL) / Magnetic Susceptibility of Magnetite under Hydrostatic Pressure, and Implications for Tectonomagnetism (A.A. NULMAN, V.A. SHAPIRO, S.I. MAKSIMOVSKIKH, N.A. IVANOV, J. KIM, and R.S. CARMICHAEL) / Effect of Uniaxial Stress upon Remanent

Magnetization: Stress Cycling and Domain State Dependence (J. Revol, R. Day, and M. Fuller) / On the Measurement of Stress Sensitivity of NRM Using a Cryogenic Magnetometer (T.L. Henyey, S.J. Pike, and D.F. Palmer)

AEPS Vol. 6

Hard cover edition to Journal of Physics of the Earth Vol. 26, Supplement, 1978 (Not included in regular issues)

Proceedings of the International Conference on Geodynamics of the Western Pacific-Indonesian Region, March 1978, Tokyo

GEODYNAMICS OF THE WESTERN PACIFIC

Edited by S. Uyeda, R.W. Murphy, and K. Kobayashi

Contents Plate Tectonic Evolution of North Pacific Rim (W.R. Dickinson) / Speculations on Mountain Building and the Lost Pacific Continent (A. Nur and Z. Ben-Avraham) / Benioff Zones, Absolute Motion and Interarc Basin (F.T. Wu) / Oceanic Crust in the Dynamics of Plate Motion and Back-Arc Spreading (Y. Ida) / Basic Types of Internal Deformation of the Continental Plate at Arc-Arc Junctions (K. Shimazaki, T. Kato, and K. Yamashina) / Fault Patterns in Outer Trench Walls and Their Tectonic Significance (G.M. Jones, T.W.C. Hilde, G.F. Sharman, and D.C. Agnew) / Motion of the Pacific Plate and Formation of Marginal Basins: Asymmetric Flow Induction (R.C. Bostrom) / The Relationship between Volcanic Island Genesis and the Indo-Australian Pacific Plate Margins in the Eastern Outer Islands, Solomon Islands, South-West Pacific (G. Wyn Hughes) / Upper Mantle Velocity Structure in the New Hebrides Island Arc Region (K.L. Kaila and V.G. Krishna) / Upper Mantle Velocity Structure in the Tonga-Kermadec Island Arc Region (K.L. Kaila and V.G. Krishna) / Morphology and Structure of the Southern Part of the New Hebrides Island Arc System (J. Daniel) / Paleomagnetic Evidence for the Rotation of Seram, Indonesia (N.S. Haile) / A Late Miocene K-Ar Age for the Lavas of Pulau Kelang, Seram, Indonesia (R.D. Beckinsale and S. Nakapadungrat) / A Survey of Paleomagnetic Data on Mexico (S. Pal) / Southeast Asian Tin Granitoids of Contrasting Tectonic Setting (C.S. Hutchison) / Seismicity, Gravity and Tectonics in the Andaman Sea (R.K. Verma, M. Mukhopadhyay, and N.C. Bhuin) / Focal Mechanisms and Tectonics in the Taiwan-Philippine Region (T. Seno and K. Kurita) / Recent Tectonics of Taiwan (F.T. Wu) / Tectonics of the Ryukyu Island Arc (K. Kizaki) / Explosion Seismic Studies in South Kyushu Especially around the Sakurajima Volcano (K, Ono, K. Ito, I. Hasegawa, K. Ichikawa, S. Iizuka, T. Kakuta, and H. Suzuki) / Two Types of Accretionary Fold Belts in Central Japan (Y. Ogawa and K. Horiuchi) / Permain and Triassic Sedimentary History of the Honshu Geosyncline in the Tamba Belt, Southwest Japan (D. Shimizu, N. Imoto, and M. Musashino) / Thermal Structure of the Sanbagawa Metamorphic Belt in Central Shikoku (S. Banno, T. Higashino, M. Otsuki, T. Itaya, and T. Nakajima) / Shimanto Geosyncline and Kuroshio Paleoland (T. Harata, K. Hisatomi, F. Kumon, K. Nakazawa, M. Tateishi, H. Suzuki, and T. Tokuoka) / Magnetic Stratigraphy of the Japanese Neogene and the Development of the Island Arcs of Japan (N. Niitsuma) / Regional Characteristics and Their Geodynamic Implications of Late Quaternary Tectonic Movement Deduced from Deformed Former Shorelines in Japan (Y. Ota and T. Yoshikawa) / Magnetic Anomalies and Tectonic Evolution of the Shikoku Inter-Arc Basin (K. Kobayashi and M. Nakada) / A Compilation of Magnetic Data in the Northwestern Pacific and in the North Philippine Sea (N. Isezaki and H. Miki) / Collision of the Izu-Bonin Arc with Central Honshu: Cenozoic Tectonics of the Fossa Magna, Japan (T. Matsuda) / Flow under the Island Arc of Japan and Lateral Variation of Magma Chemistry of Island Arc Volcanoes (M. Toriumi) / Seismic Activity and Pore Pressures across Island Arcs of Japan (N. Fujii and K. Kurita) / Aseismic Belt along the Frontal Arc and Plate Subduction in Japan (K. Yamashina, K. Shimazaki, and T. Kato) / Tsunamicity of Sanriku Depends on Subduction Tectonics (Wm. M. Adams) / Seismic Studies of the Upper Mantle beneath the Arc-Junction at Hokkaido: Folded Structure of Intermediate-Depth Seismic Zone and Altenuation of Seismic Waves (T. Moriya) / Sedimentary Patterns in Apparent Back-Arc Basins: A Case Study of the Neogene Sequence in Northwestern Hokkaido, Japan (H. Okada) / Velocity Anisotropy in the Sea of Japan as Revealed by Big Explosions (H. Okada, T. Moriya, T. Masuda, T. Hasegawa, S. Asano, K. Kasahara, A. Ikami, H. Aoki, Y. Sasaki, N. Hurukawa, and K. Matsumura) / Geodynamics of the North-Eastern Asia in Mesozoic and Cenozoic Time and the Nature of Volcanic Belts (L.M. Parfenov, I.P. Voinova, B.A. Natal'in, and D.F. Semenov) / The Crustal Structure and Origin of the Basins of Japan Sea and Some Other Seas of the Circum-Pacific Mobile Belt (P.N. Kropotkin) / Major Strike-Slip Faults and Their Bearing on Spreading in the Japan Sea (K. Otsuki and M. Ehiro) / Significant Eruptive Activities Related to Large Interplate Earthquakes in the Northwestern Pacific Margin (M. Kimura) / A Mechanism to Explain the Earthquakes around Japan by the Process of Partial Melting (M. Hayakawa and S. Iizuka) / The Formation of Intermediate and Deep Earthquake Zone in Relation to the Geologic Development of East Asia since Mesozoic (Y. Suzuki, K. Kodama, and T. Mitsunashi) / Subject Index / Geographical Index

AEPS Vol. 7

Hard cover edition to Journal of Geomagnetism and Geoelectricity Vol. 31, No. 3, 1979

Proceedings of IAGA/IAMAP Joint Assembly August 1977, Seattle, Washington

ELECTRIC CURRENT AND ATMOSPHERIC MOTION

Edited by S. Kato and R.G. Roper

Contents Electrodynamics of the Ionosphere from Incoherent Scatter: A Review (M. Blanc) / Long-Period Waves in Mesospheric Winds at Saskatoon (52°N) (A.D. Belmont and G.D. Nastrom) / Solar Tidal Wind Structures and the E-Region Dynamo (J.M. Forbes and H.B. Garrett) / Dynamics of Severe Storms through the Study of Thermospheric-Tropospheric Coupling (R.J. Hung and R.E. Smith) / Coordinated Measurements of E-Layer Drifts (E.S. Kazimirovsky and V.D. Kokourov) / On an Origin of Ultra Long Period (Several Days) of Geomagnetic Fluctuations (T. Kitamura) / IMF and Lower Thermospheric Currents and Motions: A Review (S. Matsushita) / Abnormal Features of the Regular Daily Variation S_R (P.N. Mayaud) / Rocket Measurements of Annual Mean Prevailing, Diurnal and Semi-Diurnal Winds in the Lower Thermosphere at Mid-Latitudes (D. Rees) / Mid-Latitude Winds and Electric Fields in the Lower Thermosphere and Their Relationship with the Global S_q Ionospheric Current System (D. Rees) / Ionospheric Wind Dynamo Theory: A Review (A.D. Richmond) / The Quiet-time Equatorial Electrojet and Counter-Electrojet (R.T. Marriott, A.D. Richmond, and S.V. Venkateswaran) / Equatorial Electrojet and S_q Current System —Part I (J. W. MacDougall) / Equatorial Electrojet and S_q Current System—Part II (J.W. MacDougall) / Results from *in situ* Measurements of Ionospheric Currents in the Equatorial Region—I (S. Sampath and T.S.G. Sastry) / Depth of Non-Conducting Layer in the Indian Ocean Region around Thumba, Derived from *in situ* investigations of Equatorial Electrojet— II (S. Sampath and T.S.G. Sastry) / AC Electric Fields Associated with the Plasma Instabilities in the Equatorial Electrojet—III (S. Sampath and T.S.G. Sastry) / Electric Potential

Difference between Conjugate Points in Middle Latitudes Caused by Asymmetric Dynamo in the Ionosphere (N. Fukushima) / Results of Wind Velocity Measurements at Middle and High Latitudes by the Meteor Radar Method (I.A. Lysenko, A.D. Orlyansky, and Yu. I. Portnyagin) / A Comparison between Radio Meteor and Airglow Winds (G. Hernandez and R.G. Roper) / Subject Index

AEPS Vol. 8

Hard cover edition to Journal of Physics of the Earth Vol. 27, Supplement, 1979 (Not included in regular issues)

STRUCTURE OF TRANSITION ZONE

Edited by S. Asano

Contents Crust and Upper Mantle Structure beneath Northeastern Honshu, Japan as Derived from Explosion Seismic Observations (S. Asano, H. Okada, T. Yoshii, K. Yamamoto, T. Hasegawa, K. Ito, S. Suzuki, A. Ikami, and K. Hamada) / Regionality of the Upper Mantle around Northeastern Japan as Revealed by Big Explosions at Sea. I. SEIHA-1 Explosion Experiment (H. Okada, S. Asano, T. Yoshii, A. Ikami, S. Suzuki, T. Hasegawa, K. Yamamoto, K. Ito, and K. Hamada) / On the Junction Character of the Continental and the Oceanic Lithospheric Blocks in the Kamchatka Region (G.I. Anosov, S.K. Bikkenina, V.I. Fedorchenko, A.A. Popov, K.F. Sergeev, and V.K. Utnasin) / New Evidences of the Discontinuous Structure of the Descending Lithosphere as Revealed by *ScSp* Phase (Hm. Okada)/ A Block Velocity Model of the Focal Zone and Adjacent Mantle in the Kurile-Japan Region (R.Z. Tarakanov) / Geological Structure of the Southwestern Okhotsk Sea Area (S.L. Soloviev, M.L. Krasny, O.A. Melnikov, Yu.A. Pavlov, E.I. Popov, S.S. Snegovskoy, I.K. Tuenov, and B.I. Vasiliev) / Structure and Geological Nature of the Kuril Abyssal Basin in the Okhotsk Sea (I.K. Tuezov, B.I. Vasiliev, M.L. Krasny, Yu.A. Pavlov, and E.I. Popov) / The Daito Ridge Group and the Kyushu-Palau Ridge—with Special Reference to the Tectonics of the Philippine Sea— (T. Shiki, Y. Misawa, and I. Konda) / Heat Flow in the Hokkaido-Okhotsk Region and Its Tectonic Implications (S. Ehara) / Geothermal Investigations Carried out in the Northwestern Sector of the Pacific Mobile Belt (O.V. Veselov, N.A. Volkova, V.V. Soinov, and Y.D. Eremin) / Deep Electric Conductivity Study in the Asia-to-Pacific Transition Zone (L.L. Vanyan, V.V. Boretz, A.M. Lyapishev, B.E. Marderfeld, A.V. Rodionov, and V.N. Verkhovsky) / Electrical Conductivity Structure beneath the Japan Island Arc by Geomagnetic Induction Study (J. Miyakoshi) / Geomagnetic Anomalies of the Japan Sea by Japanese and Russian Magnetic Data (N. Isezaki, M. Yasui, and S. Uyeda) / Recent Crustal Movements in Primorie and Sakhalin (V.K. Zakharov, B.A. Belousov, V.P. Semakin, N.F. Vasilenko, and G.G. Yakushko) / Subject Index

AEPS Vol. 9

Hard cover edition to Journal of Geomagnetism and Geoelectricity Vol. 32, Supplement I, 1980 (Not included in regular issues)

ELECTROMAGNETIC INDUCTION IN THE EARTH AND MOON

Edited by U. Schmucker

Contents Observation of Very Low Frequency Electromagnetic Signals in the Ocean (J. H. Filloux) / Atlantic Lithosphere Sounding (C. S. Cox, J. H. Filloux, D. I. Gough, J. C. Larsen,

K. A. POEHLS, R. P. VON HERZEN, and R. WINTER) / North Pacific Magnetotelluric Experiments (J. H. FILLOUX) / Electromagnetic Induction in the Vancouver Island Region (W. NIENABER, H. W. DOSSO, L. K. LAW, F. W. JONES, and V. RAMASWAMY) / Induction in Arbitrarily Shaped Oceans II: Edge Correction for the Case of Infinite Conductivity (R. C. HEWSON-BROWNE and P. C. KENDALL) / The Effect of a Simple Model of the Pacific Ocean on S_q Variations (B. A. HOBBS and G. J. K. DAWES) / Electromagnetic Induction at a Model Ocean Coast (G. FISCHER, P.-A. SCHNEGG, and K. D. USADEL) / Diakoptic Solution of Induction Problems (C. R. BREWITT-TAYLOR and P. B. JOHNS) / Induction by S_q (W. D. PARKINSON) / Electromagnetic Response Functions from Interrupted and Noisy Data (J. C. LARSEN) / Deep Conductivity Distribution on the Russian Platform from the Results of Combined Magnetotelluric and Global Magneto-variational Data Interpretation (A. A. KOVTUN and L. N. POROKHOVA) / Connection between the Electric Conductivity Increase due to Phase Transition and Heat Flow (A. ÁDÁM) / Geomagnetic Variations Behavior in Central Europe (I. I. ROKITYANSKY) / Geomagnetic Sounding of an Ancient Plate Margin in the Canadian Appalachians (J. A. WRIGHT and N. A. COCHRANE)/ Magnetovariational and Magnetotelluric Investigations in S. Scotland (V. R. S. HUTTON and A. G. JONES) / An Analogue Model Study of Ocean-Wave Induced Magnetic Field Variations Near a Coastline (T. MILES and H. W. DOSSO) / Long Period Variations of the Geomagnetic Field and Inferences about the Deep Electric Conductivity of the Earth (A. M. IŞIKARA) / Inverse Magnetotelluric Problem for Sounding Curves from Czechoslovak Localities (K. PĚČ, J. PĚČOVÁ, and O. PRAUS) / Remarks on Spatial Distribution of Long Period Variations in the Geomagnetic Field over European Area (J. PĚČOVÁ, K. PĚČ, and O. PRAUS) / An Interpretation of the Induction Arrows at Indian Stations (B. J. SRIVASTAVA and H. ABBAS)

AEPS Vol. 10

Hard cover edition to Journal of Geomagnetism and Geoelectricity Vol. 32, Supplement III, 1980 (Not included in regular issues)

Proceedings of IUGG Symposium "Global Reconstruction and the Geomagnetic Field during the Palaeozoic" Canberra, 1979, December

GLOBAL RECONSTRUCTION AND THE GEOMAGNETIC FIELD DURING THE PALAEOZOIC

Edited by M. W. MCELHINNY, A. N. KHRAMOV, M. OZIMA, and D. A. VALENCIO

Contents Magnetostratigraphy in the Sydney Basin, Southeastern Australia (B. J. J. EMBLETON and K. L. MCDONNELL) / Early Palaeozoic Palaeomagnetism in South East Australia (B. R. GOLEBY) / Palaeomagnetism and Reconstruction of Palaeogeographic Positions of the Siberian and Russian Plates during the Late Proterozoic and Palaeozoic (A. N. KHRAMOV and V. P. RODIONOV) / The Concept of a Mobile Pangea and the Continuity of Continental Drift (P. MOREL and E. IRVING) / Model of the World Plate Tectonics since the Early Palaeozoic (V. BUCHA) / A Computer Animation of Continental Drift (C. R. SCOTESE, S. SNELSON, W. C. ROSS, and L. P. DODGE) / Palaeomagnetism of Lower Ordovician and Upper Precambrian Rocks from Argentina (D. A. VALENCIO) / Magnetism of the Mid-Ordovician Tramore Volcanics, SE Ireland, and the Question of a Wide Proto-Atlantic Ocean (E. R. DEUTSCH) / The Geomagnetic Field during Palaeozoic Time (A. N. KHRAMOV and V. P. RODIONOV) / Apparent Polar Wander Path of Central Iran and Its Geotectonic Interpretation (H. C. SOFFEL and H. G. FÖRSTER) / Reversals and Excursions of the Geomagnetic Field as Defined by Palaeomagnetic Data from Upper Palaeozoic-Lower Mesozoic Sediments and Igneous Rocks from Argentina (D. A. VALENCIO)